Lecture Notes in Control and Information Sciences

Edited by M. Thoma and A. Wyner

For information about Vols. 1–61 please contact your bookseller or Springer-Verlag.

Lecture Notes in Control and Information Sciences

Edited by M. Thoma and A. Wyner

132

W.-Y. Ng

Interactive Multi-Objective Programming as a Framework for Computer-Aided Control System Design

Springer-Verlag
Berlin Heidelberg GmbH

Author
Wai-Yin Ng
Dept. of Information Engineering
The Chinese University of Hong Kong
Shatin, N. T.
Hong Kong

ISBN 978-3-540-51504-3   ISBN 978-3-540-48183-6 (eBook)
DOI 10.1007/978-3-540-48183-6

2161/3020-543210   Printed on acid-free paper.

To Jessica

"There are many contradictions in the process of the development of a complex thing, and one of them is necessarily the principal contradiction whose existence and development determine or influence the development and existence of the other contradictions ... but this situation is not static; the principal and non-principal aspects of a contradiction transform themselves into each other ..."

Mao Zedong

Preface

This monograph reports the development of an approach to computer-aided control system design (CACSD). Of primary concern is the quality of co-operation between the designer and his computer. In order to achieve an effective co-operation, we propose a conceptual view of the CACSD problem to the designer, and develop a framework for the computer. The conceptual view is named Generalized Co-operative Search (GCS), which suggests the designer to treat a CACSD problem as a search problem, in which he and the computer co-operate to locate satisfactory designs among sets of candidates. The framework is Interactive Multi-Objective Programming (IMOP), which defines the level of abstraction as well as the organization of the design facilities in the computer.

While IMOP methods and control system design methods provide readily usable tools organized by the framework, a design strategy is developed to guide the designer in their effective use. The strategy is a two-level plan of the design process which helps the designer to conduct his search problems as ones of IMOP. He constructs parametrized sets of candidate designs using the design methods (innovates in level I) and matches his design wishes with the design possibilities amongst these candidates using the IMOP methods (trade-offs in level II).

The advantages of this approach are (1) a proper emphasis is put on an effective designer-computer co-operation, the designer is supported rather that forced to follow any planned course of actions;

(2) a uniform trade-off among design objectives receive their long due attention and (3) it is pluralistic and different design methods may be combined for their respective strengths.

The work conducted is likely to be the first integrated approach to designing control systems by search, and may even be the first which supports the principal design stages (formulation, generation and evaluation) in a co-ordinated as well as most general manner. The proposed approach has a promising prospect of being one for the general areas of computer-aided design in engineering as well as decision support system in modern management.

Acknowledgement

It is a pleasure to express my thanks to Dr. Jan M. Maciejowski for his constant help and guidance as well as Prof. A.G.J. MacFarlane, with whom I have had many enlightening discussions. Their recommendation of my going to the DFVLR in West Germany is the turning point of my course to the design by search approach.

This work is done during my stay in Cambridge, England, which has been made possible by the Benefactor's Studentship of St. John's College as well as the Overseas Research Scholarship. This research has received financial support from SERC, grant no. GR/D/60782. My sincere thanks go also to Mr. David K.P. Li and the Friends of Cambridge in Hong Kong who first sent me to Cambridge in 1982.

I am grateful to the people with whom I have had many useful exchanges, namely, Dr. V. Zakian of UMIST, U.K., Dr. R. Steinhauser and Prof. G. Grubel of DFVLR, West Germany, Phil Smith of the Royal Aircraft Establishment, Bedford, and Jean-Marc Boyle, Lai-wan Chan, James Lam, Raimund Ober, Peter Phaal and Neil Piercy, all of CUED. I also thank Barry Ko for helping me with the graphs and Miss Karen Cheung for preparing the manuscript.

I thank Dr. V. Zakian for arranging access to his CRITERIA package in UMIST, the RAE for providing the linear models of the V/STOL aircraft used in chapter 9, and Dr. H. Michalska of the Imperial College, London, for the use of DELIGHT.

CONTENTS

NOTATION

Unless otherwise stated, the following notation will be adopted :

$a . * b$	element-by-element multiplication of vectors a and b
$\|y\|$, $\|y\|_p$	a norm and an l_p-norm of vector y respectively
$Im(\omega)$	the imaginary part of ω
$Re(\omega)$	the real part of ω
$i \mathcal{R} j$	indices i and j are corresponding; section 7.2
$Q_{l_1} \mathcal{R} Q_{l_2}$	index groups $l1$ and $l2$ are corresponding; section 7.2
$i \mathcal{C} j$	indices i and j are conflicting; section 7.2
$Q_{l_1} \mathcal{C} Q_{l_2}$	index groups $l1$ and $l2$ are conflicting; section 7.2
$i \mathcal{U} j$	indices i and j are unrelated; section 7.2
$Q_{l_1} \mathcal{U} Q_{l_2}$	index groups $l1$ and $l2$ are unrelated; section 7.2
$c \in \Re^q$	$= (c_1, c_2, \ldots, c_q)^T$
	a vector of performance index bounds
$\hat{f} : \Re^n \mapsto \Re^q$	$\hat{f}(x) = (\hat{f}_1(x), \hat{f}_2, (x), \ldots, \hat{f}_q(x))^T$
	vector aggregate performance index function; section 5.3.2
$\hat{f}_i^g$, $\hat{f}_i^b$	assigned good and bad values for aggregate performance index $\hat{i}$;
	section 6.1
$\bar{f}_i : \Re^q \mapsto \Re$	a scaled aggregate performance index; section 6.1
$f : \Re^n \mapsto \Re^q$	$f(x) = (f_1(x), f_2, (x), \ldots, f_q(x))^T$
	vector performance index function
$F : \Re^q \mapsto \Re$	a scalarizing function
$F_w : \Re^q \mapsto \Re$	a weighted sum scalarizing function; section 3.7.1
$F_\epsilon : \Re^q \mapsto \Re$	an ϵ-constrained scalarizing function; section 3.7.1

$F_{d_1} : \Re^q \mapsto \Re$ a weighted distance norm scalarizing function with a vector of non-negative weights as the scalarizing parameters; section 3.7.1

$F_{d_2} : \Re^q \mapsto \Re$ a weighted distance norm scalarizing function with a reference point as the vector of scalarizing parameters; section 3.7.1

$F_a : \Re^q \mapsto \Re$ an achievement function; section 3.7.1

$g_i(x, z) \leq 0, \quad \forall z \in Z_i$

a functional constraint

$g_{l_1 l_2}$ the median linkage between index groups Q_{l_1} and Q_{l_2}; section 7.3

$G(s_r)$ $= \{g_{l_1 l_2}\} \in \Re^{m(s_r) \times m(s_r)}$

a median linkage matrix for index groups formed with threshold of correspondence s_r; section 7.3

j^* the most active hard constraint; section 6.1.3

n the dimension of the design parameter space

P $= \{Q_l, \ l \in M = \{1, 2, \ldots, m\}\}$

a partition of the index set Q

$\mathcal{P}(M)$ $= ([1], [2], \ldots, [m])$

a permutation of the elements of a set $M = \{1, 2, \ldots, m\}$

$\hat{Q}$ $= \{1, 2, \ldots, \hat{q}\}$

index set (aggregate performance indices); section 6.1

$\hat{q}$ the dimension of the aggregate performance index space

Q $= \{1, 2, \ldots, q\}$

index set (performance indices); section 3.2

q the dimension of the performance index space

$\tilde{Q}$ $= \{i_l, \ l \in M\}$

the set of group active members

$Q_s \subset Q$ the set of performance indices which bounds are satisfied by the initial point in an optimization search; section 6.2.2

$Q_u \subset Q$ the set of performance indices which bounds are not satisfied by the initial point in an optimization search; section 6.2.2

$\Re^m$ m-dimensional Euclidean Space

$\Re^q_+$ $= \{y \in \Re^q : y_i \geq 0 \quad \forall i = 1, 2, \ldots, q\}$

non-negative orthant of $\Re^q$

$\dot{\Re}^q_+$ $= \{y \in \Re^q : y_i > 0 \quad \forall i = 1, 2, \ldots, q\}$

positive orthant of $\Re^q$

s_{ij} $(1 \geq s_{ij} \geq -1)$

the sample rank correlation between performance indices i and j; section 7.2

$S \in \Re^{q \times q}$ $= \{s_{ij}\}$

sample rank correlation matrix of performance indices; section 7.2

s_c an assigned threshold of conflict; section 7.3

s_r an assigned threshold of correspondence; section 7.3

$U : \Re^q \mapsto \Re$ utility function

$X \subset \Re^n$ the set of feasible design parameter vectors

$x \in X$ $= (x_1, x_2, \ldots, x_n)^T$

a feasible design parameter vector

$X^* \subset X$ efficient set; section 3.3

$x^* \in X^*$ an efficient point in X

X^S $= \{x_h \in \Re^n, h \in P^S = \{1, 2, \ldots, p\}\}$

sample efficient subset; section 7.1

Y the set of feasible performance index vectors

$y \in Y$ $= (y_1, y_2, \ldots, y_q)^T = f(x), \quad \exists x \in X$

a feasible performance index vector

$Y^* \subset Y$ efficient frontier; section 3.3

$y^* \in Y^*$ an efficient performance index vector in Y

y^{**} $= (y_1^{**}, y_2^{**}, \ldots, y_q^{**})^T$

the ideal point; section 3.7.1

$\bar{y} \in \Re^q$ a reference point in $\Re^q$, the space of performance index vectors

Y^S $= \left\{ y_h \in \Re^q, h \in P^S \right\}$

sample efficient frontier; section 7.1

$\bar{Y}(A) \subset Y$ $= \{ y(\alpha) : \alpha \in A \}$

the image of the set of all admissible vectors of scalarizing

parameters A; section 3.3

Z a matrix representation of a simplex polytope, with columns of

position vectors of the polytope's apexes; section 6.1.1

List of Abbreviations

Acronyms

AN   the ANalyst in IMOP; section 3.1

CACSD   Computer-Aided Control System Design

CAD   Computer-Aided Design

DM   the Decision Maker in IMOP; section 3.1

GCS   Generalized Co-operative Search

IAE   Integral Absolute Error

IMOP   Interactive Multi-Objective Programming

LQR   Linear Quadratic Regulator

NPTG   Nuclear-Powered Turbo-Generator; section 9.1

V/STOL   Vertical/Short Take-Off and Landing aircraft; section 9.2

Working Assumptions (section 5.1)

[WA1]   All Efficiency

[WA2]   Efficient Trajectory

Design Principles (section 5.2)

[DP1]   Pluralistic Use of Design Methods

[DP2]   Uniform Trade-offs and Evaluations

[DP3]   Rapid Prototyping/Empirical Learning

[DP4]   Design Granularity

[DP5]   A Priori Search Control

[DP6]   Open Design

Graph Types

[P1]   e.g. figure 6.6 (section 6.1.3)

[P2]   e.g. figure 6.7 (section 6.1.3)

[P3]   e.g. figure 6.8 (section 6.1.3)

[P4]   e.g. figure 6.9 (section 6.1.3)

[P5]   e.g. figure 6.12 (section 6.2.2)

[A2P1]   e.g. figure 7.10 (section 7.3)

[A2P2]   e.g. figure 7.11 (section 7.3)

CHAPTER 1

INTRODUCTION

This thesis reports the development of an approach to computer-aided control system design (CACSD), based on interactive multi-objective programming (IMOP).

1.1 The Control Problem

Control problems occur frequently in engineering and management, and indeed, whenever a purposeful agent (in most cases human) interacts with a dynamic environment. A person having a shower has the problem of regulating the temperature and flow rate of the water. A pilot has a problem of controlling his aircraft for his mission. The operators of a nuclear plant have to control the various critical parameters of the fuel core for safety and energy output. The generic problem they all face is the control problem. In each case, there is a **plant**, which is the system to be controlled, and a **controller**, which represents the purposeful agent. Feedback is a central concept of most control strategies (fig. 1.1). The plant outputs are monitored by sensors which notify the controller who compares them with the reference levels, which represent the desires of the agent. Any deviation results in actions which alter the plant inputs through actuators. Sensors and actuators are both transducers which translate signals of different forms. Different controllers may be used on the same plant at different operation modes or for different purposes. The collection of controllers constitute the control system.

However, control problems are often dealt with only implicitly. No separate controller is built and the purposeful agent directly handles the plant. He is the

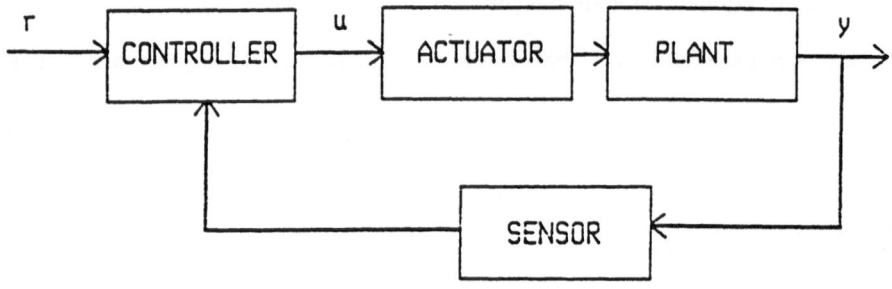

r   reference levels
u   actuator inputs
y   plant outputs

1.1 Feedback Control

controller and manual control is exercised. In cases when control is tackled as a problem in itself with the plant given and a control system designed and implemented to represent the agent, automatic control is exercised. CACSD is the area which is concerned with *the efficient employment of computers in the design of control systems.* With the advances in the critical technologies of control engineering (e.g. transducer, microprocessor) and the rising demand for higher precision in control, and the possibility of controlling complex systems, automatic control has become widely practised, especially among the high technologies, e.g. aerospace, robotics, high-performance chemical plants, etc. CACSD has therefore become a critical technology for the success of such practical applications.

Practical design is often complicated by the many objectives in the specifications. This cannot be over-emphasized in control system design. A good control system is expected to stabilise the plant, improve the transient responses (when reference levels change) and the steady- state characteristics (when reference levels are locked), provide disturbance rejection, and decrease sensitivity to parameter variations [Van 86]. Apart from these control-theoretic objectives, there are considerations of the various costs (control cost, implementation cost, operating and maintenance costs), complexity, implementation, handling qualities, safety, etc.

1.2 Designer-Computer Interaction (Fig. 1.2)

When a designer is faced with a design problem, he will develop a conceptual view which is a particular way of looking at it. The view depends on the problem itself and is affected by the design facilities available to him. Subsequently, his plan to resolving the problem will originate from it.

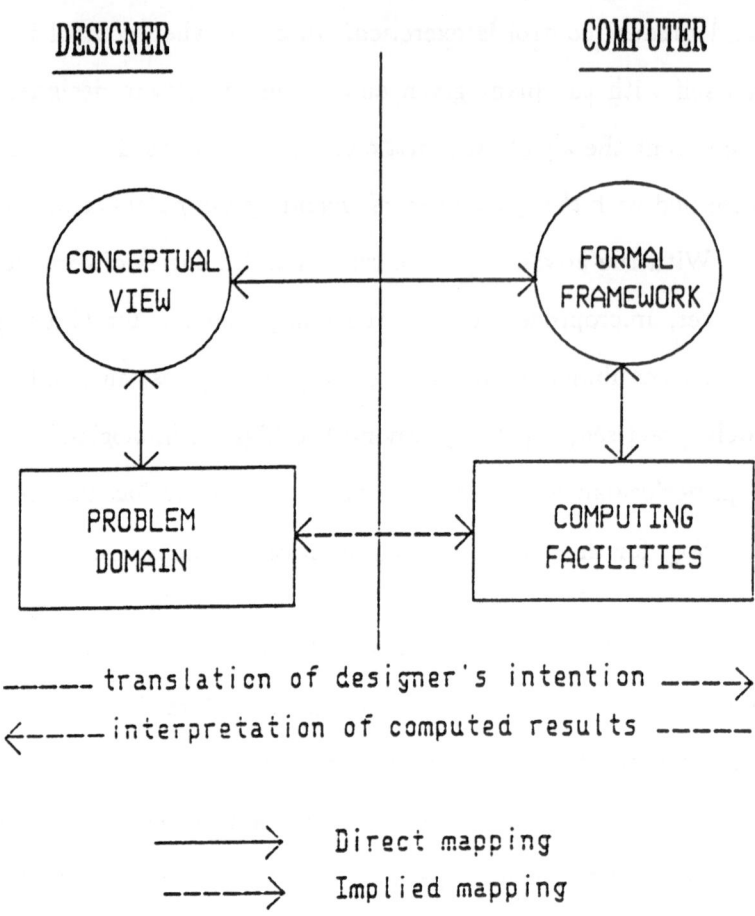

DESIGNER   COMPUTER

translation of designer's intention ⎯⎯⟶
⟵⎯⎯ interpretation of computed results ⎯⎯

⎯⎯⟶ Direct mapping
------⟶ Implied mapping

1.2 Design-Computer Interaction

In CACSD, a designer interacts with his computer for the design facilities. He translates his intention into computing procedures and subsequently obtains better understanding by interpreting the computed results. This interaction is strongly affected by the organization of the design facilities in the computer. We shall call this organization the computer's **formal framework**. It is formal in the sense that it can be fully expressed unambiguously, and is therefore implementable as software.

The framework is of primary importance to the quality of interaction between the designer and the computer. If the framework bears a good correspondence with the designer's conceptual view, this will result in a more direct implied mapping between the design problem and the computing facilities, as shown in fig. 1.2. This greatly reduces the cognitive load imparted on the designer in translating intention and interpreting results. Cognitive compatibility is said to have been achieved and a good quality of interaction between the designer and the computer may be expected [Nor 86]. Otherwise, the design facilities will be difficult to use and attention will be drawn away from the design problem itself.

1.3 Our CACSD Approach (Fig. 1.3)

Our belief is that the quality of interaction between the designer and his computer is of primary importance to the success of a CACSD approach. To achieve good interaction, we need to establish a good correspondence between the designer's conceptual view and the computer's formal framework. To achieve this, we explicitly developed a conceptual view called the **generalized co-operative search (GCS)** for the designer and identified **interactive multi-objective programming (IMOP)** as the corresponding formal framework for the computer.

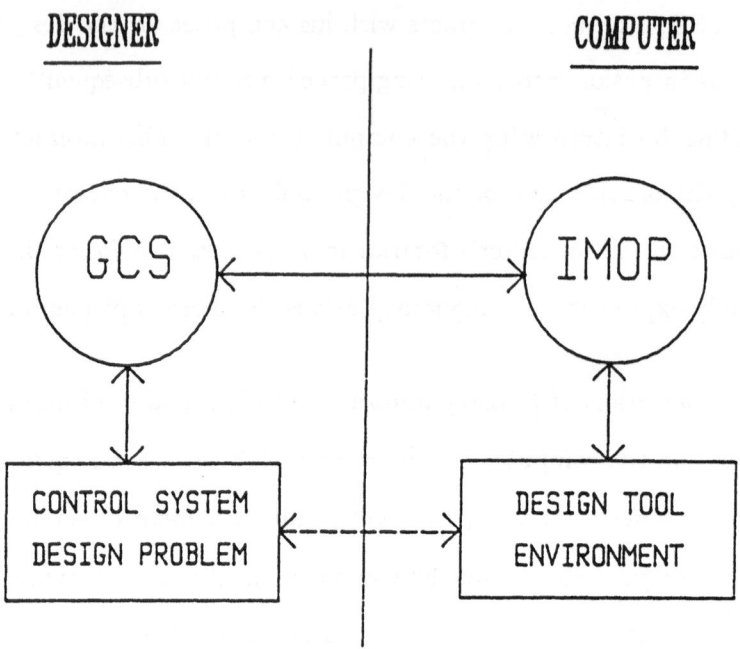

DESIGNER   COMPUTER

GCS   Generalized Co-operative Search
IMOP   Interactive Multi-Objective Programming

1.3 Proposed Scheme for Design-Computer Interaction for CACSD

Having established a good quality of designer-computer interaction, we then developed a strategy for the designer which is a plan to employ this interaction in accessing the design facilities (organized by IMOP) to resolve the design problem (seen as a GCS problem). A set of design principles is identified which are used as the overall objectives of the strategy, with strong emphasis on improving the designer's understanding. The strategy itself consists of two levels. Candidate designs are innovated and constructed in level I with the help of various design methods. In level II, trade-offs among the various design objectives are handled when the candidates from level I are parametrized and numerical search is conducted in its neighbourhood in the resulting parameter space.

The novelty of our approach lies in (1) the explicit construction of a conceptual view and a formal framework to achieve a good quality of designer-computer interaction and (2) the emphasis on an interactive approach to trade-offs as a means to match designer's wishes with design possibilities.

We feel that the improvement of the man-machine interface is often done in a manner too empirical and ad hoc. Our approach is an attempt to be more normative, especially in the construction of the framework, by developing an explicit conceptual view for the designer to adopt.

We also find that trade-offs have been a very much neglected aspect of current CACSD approaches. Although good design methods always enable the designer to handle some well-known trade-offs, e.g. the LQR method which handles the trade-off between performance and control cost with cost matrices, the objectives being traded-off are *prescribed by the method*. However, in practice, the designer is always handling many objectives *determined by the design specifications*, which may include some which are non-control-theoretic. A uniform trade-off procedure is lacking in CACSD, while in practice, design often proceeds as progressive resolution of conflicts among design objectives by trade-offs. Indeed, design can be seen as iterations of trade-offs, which are attempts to *match designer's wishes with design possibilities.*

In our strategy, level I is an initialization step for the subsequent transcription of the CACSD problem as one of IMOP. Level II then handles the iterations of trade-offs based on the interactive approach of IMOP. The strategy is carried out with the help of a set of procedures developed within the framework of IMOP. These procedures include numerical search techniques with graphical monitors, a data compression and description technique to help the designer in discovering and comprehending the relationship among objectives, and interactive procedures as aids to making trade-off decisions and guiding the search.

1.4 Implementation

To implement the design facilities within the proposed framework, we are in favour of **software tool environments** [Ker 76] (as preferred to packages).

The main difference between a package and a tool environment is that a package provides a man-machine interface with the functionality in a mixed and often closed manner, closed in the sense that no additional facilities can be included naturally, while in the case of a tool environment, there is a separation between the man-machine interface and functionality. The man-machine interface is provided as a language-based interaction while the functionality is provided by a set of software tools which can be updated and organised in various ways. The designer can expand his set of tools simply by adding new functions built out of existing ones. These tools reside within an environment which provides a uniform interface through which the user employs them.

In recent years, there has been a movement away from the package approach towards the software tool environment approach in the CACSD community [Mac 84], [Wet 86]. Procedures and algorithms for design methods are then implemented as tools. The uniform interface to access these tools releases the designer from having to learn the syntax of different packages for all the facilities he requires. He may add his own tools easily if they are not readily available. A clear indication of this movement is the ever-increasing popularity of the class of CACSD softwares which are Matlab-based [Mol 81], such as Pro-Matlab [Mat 87], Program CC [Tho 85] and Ctrl-C [Lit 84].

The investigation of our design approach was carried out by building a prototype in a tool environment on a graphics engineering workstation. We chose Pro-Matlab for the extensive set of tools already available. The set of procedures required for our approach were built as a set of tools. The design of multivariable linear time-invariant (LTI) controllers is the primary problem domain addressed. Special design facilities for such problems are provided by the Control System Toolbox [Mat 87].

1.5 An Alternative View

The relevance of the work reported here can also be appreciated from the view of developments in the disciplines of CACSD and mathematical programming as induced from that in computer technology.

What digital computers offered when they first appeared was purely numerical computational power, or number crunching. The early scientific computer language FORTRAN in its original form provided nothing but a means to translate numerical algorithms to computer codes, which can then be submitted to the computer for batch execution. Mathematical programming seized the power and a whole wealth of iterative optimization algorithms for optimizing a single objective function were already well-developed by the late sixties [Lue 73]. Their success hinged on the possibility of implementing their resulting numerical algorithms. As far as control system design is concerned, this power can be used in the synthesis of controllers as closed form solutions of well-posed mathematical problems, such as optimal control problems. The boom of modern control actually relied on this computational power for the heavy computation required in the solution algorithms [Fra 87].

At the same time, computer technology was being revolutionised by advances in system software [Tan 87]. Operating systems provide an extended machine shielding the hardware from the user. Submission of computer jobs became much easier and results were obtained in minutes rather than hours. The availability of computer graphics also helped to speed up the response time by presenting results in forms more readily comprehensible to the user. Interactive computing was taking shape. This led to new development both in CACSD and mathematical programming in the seventies. There were interactive design methods in CACSD such as the Inverse Nyquist Array method [Ros 74] and the Characteristic Locus method [Mac 80].

At this time, Geoffrion et al. [Geo 72] published their paper on interactive goal programming which is an interactive method for resolving multi-objective problems employing classical mathematical optimization algorithms. This led to the birth of the field of interactive multi-objective programming which continues to be active

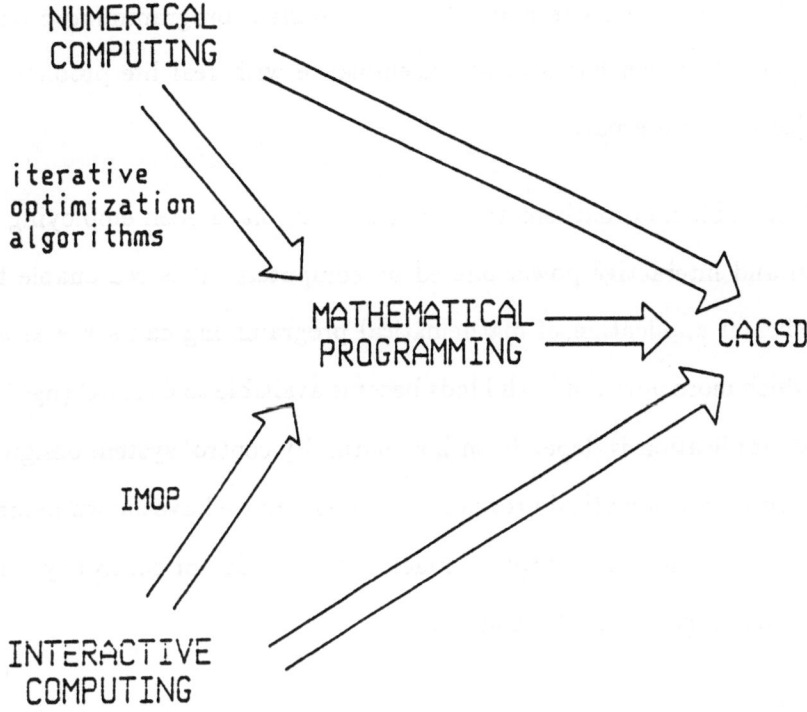

1.4 Channelling Computing Power to CACSD

till now. The interactiveness is employed to tackle a programming problem with many objectives, which has a closer resemblance with real life problems than its single-objective counterpart.

Both CACSD and mathematical programming have been employing the computational and interactive power offered by computers. It is reasonable to expect that an efficient application of mathematical programming can serve as a channel through which more power of both kinds become available to CACSD (fig. 1.4). How efficient an application is depends on how naturally control system design problem can be posed as mathematical programming ones. As we have shown before, multi-objective programming does capture much of the flavour of control system design and is an obvious candidate for such use.

1.6 Organisation (Fig. 1.5)

The organisation of the thesis is as follows.

In chapter 2, we briefly describe the various possibilities of CACSD methods. The conceptual view of CACSD as generalized co-operative search (GCS) is developed and elaborated. GCS provides a characterization of the different design methods and highlights the importance of the quality of the designer-computer interaction. A structure of GCS is then introduced as four inter-connected modules, which represent the partitioning of a CACSD problem into four sub-problems, viz. the search for candidate designs, design data management, evaluation and decision-making.

Chapter 3 elaborates the field of IMOP and features are highlighted to argue in favour of its being employed as a suitable framework for CACSD. This is done

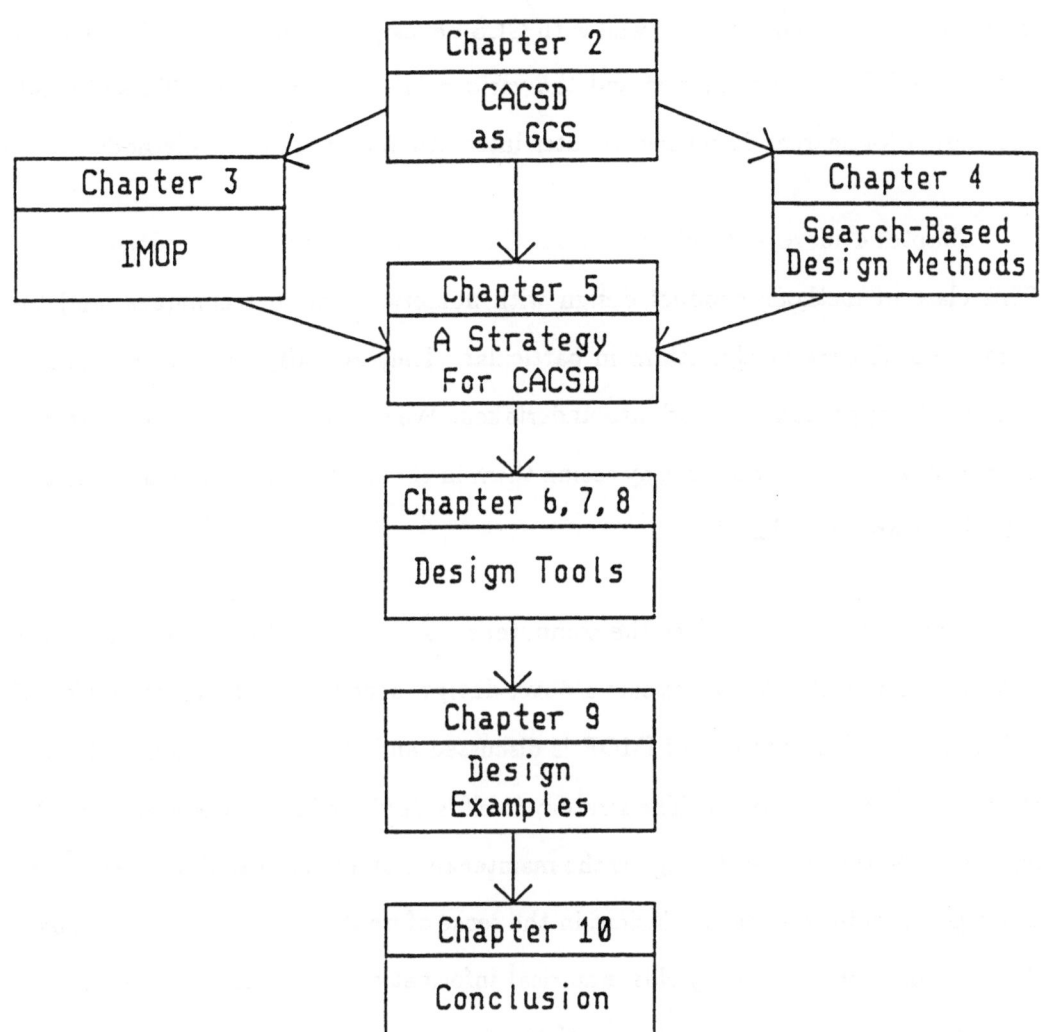

1.5 Organization of the Thesis

by showing how it corresponds well with GCS developed in chapter 2. The relevant aspects of IMOP are examined and the various methods surveyed. These include techniques for search, designer-computer interaction, and trade-off methods.

Chapter 4 surveys the class of search-based methods for control system design. This class of methods conduct design with numerical search techniques, and the iterative optimization algorithms in particular. They actually provide the inspiration for the approach this work has undertaken. We evaluate these methods as how usable they are to the designer by seeing them in the light of GCS. Useful ideas and techniques are also identified.

Having adopted IMOP as the computer's framework and derived useful ideas from the survey, chapter 5 presents a novel **design strategy**. The transcription of a CACSD problem into one of IMOP is discussed and the design principles for the strategy are then described. The strategy itself is described as a two-level plan. An important feature of the strategy is the maintenance of a set of candidate designs as a **sample efficient subset** (efficient in the sense of multi-objective programming), the interrogation of which yields empirical information to enhance the designer's understanding. The implementation of the various procedures for the strategy as tools is described in the ensuing three chapters.

Chapter 6 describes the tools for the generation of candidate designs for IMOP using iterative numerical search. Two functions of such search are identified as (1) the generation of candidates with desirable trade-offs and (2) the sampling of a set of efficient candidates. The sampling requirement points to the use of the simplex polytope method, which is a direct search method. Interestingly, due to its notoriously bad convergence property as an optimization routine, it is capable of conducting a more uniform sampling. However, reduction of the search to a linear subspace, or dimension collapsing, is a major problem of the method. Two "healing" procedures are proposed, so that the designer may monitor the degree of dimension collapsing and interrogate to execute a healing procedure to restore a full dimension search. A set of graphics monitors are also constructed to give the designer better understanding and control of the search process.

Chapter 7 describes two interactive tools for problem formulation. The designer is assisted to understand the design possibilities revealed as well as to match design wishes with such possibilities. Data visualization [Tuf 83] is the general approach and a liberal use of two-dimensional graphical displays are employed. The first tool implements a data compression and description technique to analyse relationships among design objectives. The second tool implements a novel interactive algorithm which engages the designer in a structured dialogue with the computer, which displays design specifications (which represent design wishes) against a background of the sample efficient subset (which represent revealed possibilities) in graphical forms. Such displays help the designer to appreciate any critical conflicts among the specifications, as well as possible trade-offs which may resolve them.

Chapter 8 describes the evaluation tools which interrogate the design data as captured in the sample efficient subset. We aim to support evaluations as the comparisons of a large number of candidates to complement the conventional single design evaluations. The requirement is essentially for relational database querying facilities. SEQUEL, A popular database query language is described. With simple extensions to its syntax, a language for evaluation is obtained. The language offers an efficient medium for the designer to express his intentions in evaluations by comparisons.

Chapter 9 presents two control system design examples conducted with the design approach developed. The first example is relatively straightforward to illustrate what the design process is like when the strategy is applied. The second one is a non-trivial design problem which serves to demonstrate the power and flexibility of the approach. It is shown that this approach is effective in handling many objectives. Possibilities were better understood and the designer was at a better position to make trade-off decisions.

Chapter 10 summarizes and discusses the contributions of the present work, concludes with some observations, and makes suggestions for future works.

CHAPTER 2

CACSD AS GENERALIZED CO-OPERATIVE SEARCH

2.1 Introduction

In this chapter, we first establish a general classification of CACSD methods. The conceptual view of CACSD is introduced in the subsequent section. The view is named generalized co-operative search (GCS). It communicates our concern for it to, firstly, encompass most if not all CACSD methods for a faithful representation of the practice of CACSD, and secondly, highlight the importance of co-operation between the designer and his computer.

2.2 The Classification of CACSD Methods

Broadly speaking, there are three possibilities of CACSD methods depending on the amount of analytical knowledge the designer has at his disposal [Mac 87].

2.2.1 Synthesis Methods

For CACSD problems which have exact specifications with an exactly computable answer, the solution is synthesised. The design solution is a simple function of the specifications, simple not in the sense that the function is not numerically difficult to evaluate, but that it requires no decision making from the designer during evaluation. Pole placement is a good example. For this kind of design activity, the computer simply acts as a number crunching machine, automating the synthesis procedures.

2.2.2 Procedure-based Methods

For design problems whose specifications are not exact, but nevertheless are within domains with extensive knowledge available, procedure-based methods may be developed. The designer analyses the problem and proceeds to solve it by following a series of structured procedures which provide plans of "what to examine now, what to decide, what to do next". The designer follows these procedures and manipulates a candidate design with the available operations until it meets the specifications. He makes decisions about the manipulations and even alters the specifications as he proceeds. Feedback from the procedures or immediate results are important as they guide the designer in making the decisions. Yet to obtain guidance requires the designer to interpret the feedback in the context of the problem domain *and* of the procedures. Understanding of *both* is crucial to the success of a design.

Multivariable frequency domain design methods such as the Characteristic Locus Method [Mac 80] and the Quantitative Feedback Method [Hor 79] are of this class. Very often, a subset of the procedures may be reiterated to modify the design should it be unsatisfactory. The computer acts in a closer relationship with the designer than in the case of synthesis. It may be programmed to follow the series of structured procedures with the designer. It again automates the number crunching, but in addition, it also organises information to feedback to the designer. To further computer-aid such methods points to expert systems which capture knowledge about the design methods, for example, the expert system MAID [Pan 86] which assists the designer in the execution of an organized set of multivariable frequency domain design methods.

2.2.3 Search-based Methods

The third possibility is the class of search-based design methods. It is a generate-and-test approach [Win 84] to general problem solving. Conceptually, the designer has at hand a set of controllers which can be searched through to find one which satisfies his specifications. The set is usually known by a Euclidean parametrisation. Numerical search techniques, notably iterative optimization algorithms are then applied to find controllers achieving good values of constructed objective functions. The designer has a generator which generates candidate designs from the controller set. If a satisfactory design is not found yet, the designer guides the generator in further generation of candidates. Zakian's Method of Inequalities [Zak 73] and the semi-infinite programming approach of Polak, Mayne and co-workers [Pol 84] belong to this class.

The quality of interaction is of most importance to the success of a search-based method. Unlike the procedure-based methods which tend to structure the interactions (by planning "what to see now, what to decide, what to do next"), they tend to rely on the designer for guiding the interactions. He formulates design problems by providing parametrisations and constructing objective functions. He invokes and terminates searches, examines the progress, and decides on further actions.

A fast interaction can very efficiently exploit the short-term memory of the designer, which may play an important role in such searches. A good example is the simultaneous pole placement approach using graphics animation due to Kaesbauer [Kae 86]. The designer controls up to six parameters of a controller with a steering ball while the effect on the closed loop pole positions is dynamically displayed in real time graphics. This approach has proved to be successful in practice.

2.2.4 Discussion

All three classes of methods are used in practice. The designer uses whatever is natural for his situations. However, we agree with MacFarlane et al. [Mac 87] who note :

"The most useful single characterisation of the design process is in terms of a recursively executed search procedure, drawing interactively on a rich framework of codified knowledge, accessing a wide range of tools, procedures and models in a powerful manipulative framework, and carried out in the context of a powerful and intuitively understandable conceptual framework."

The class of search-based methods best supports the design process as a whole since the concept of search, upon which they are based, is "the most useful single characterisation". The synthesis and procedure-based methods are captured within the "rich framework of codified knowledge" and implemented as lower-level operations within the design process itself. We shall develop our conceptual view of CACSD from this observation.

2.3 A Conceptual View of CACSD : Generalised Co-operative Search

This conceptual view is intended to capture, organise and clarify various concepts and is meant to be a sufficient abstraction of the design process for the designer himself. Embracing the view will keep him in sight of the highest level of the design process which he himself is involved in. Design tasks may be more readily identified and their relationship to each other better appreciated. This should encourage him to take more systematic design iterations, especially when much experimenting and trial-and-error is often present due to the lack of knowledge.

To faithfully represent the concepts of CACSD, GCS should include most, if not all, methods of design used by the designer. The concept of search is therefore most appropriate. We also see the quality of co-operation between the designer and his computer as the most crucial factor of success for any CACSD approach. We describe GCS as follows.

Firstly, CACSD is a search recursively executed within the universe of all controllers. This search is generalized in the sense that it is conducted in a variety of very different ways, among subsets of controllers within the universe. At one extreme, a particular controller may be constructed from scratch, and design proceeds as addition of modifications. This can be viewed as a search conducted along the direction of increasing complexity, within a set of controllers not explicitly known. The designer may be following a procedure-based method in this case. At another extreme, a parametrisation of a set of controllers may have been identified and a search-based method is employed to search among this explicitly known set.

The search is co-operative since there are two agents operating together, viz. the designer and his computer. Therefore, to provide any computing support for CACSD, a primary objective should be to ensure its compatibility and co-operation with the designer. A first requirement is to identify the component tasks of CACSD and then to seek an appropriate sharing of them between the designer and the computer, which we shall establish in the following two sections.

2.4 The Structure of Generalised Co-operative Search (Fig. 2.1)

We suggest in the following a structure of GCS as the co-ordination of various component tasks within CACSD. This structuring takes the object-oriented approach of modelling in which component tasks of CACSD are classified as object modules. Four kinds of modules are identified which represent the decomposition of a CACSD problem into four sub-problems.

The four kinds of modules are described in the following.

(a) Generators

Generate candidate designs on accepting guidance in the form of controller design formulations. These are intended to provide the abstraction for most if not all control system design methods. They may include a batch program which synthesizes LQR controllers, an interactive one which constructs controllers using the Characteristic Locus Method, or for search-based methods, an optimizer which searches in a defined parameter space parametrizing a certain class of controllers. In the optimizer's case, there also are intermediate designs generated along the trajectories traversed apart from the candidates. One thing these generators have in common is that the formulations they accept are better defined and narrower than the design specifications themselves. We call these the **substitute problem formulations** of the generators.

(b) Evaluators

Evaluate the candidate solutions in terms of the original design objectives. In CAD, they are interactive, with the computer supporting the evaluation by displays of design information. The designer is playing a dominant role in the evaluation. However, we distinguish an evaluator from one which evaluates intermediate designs

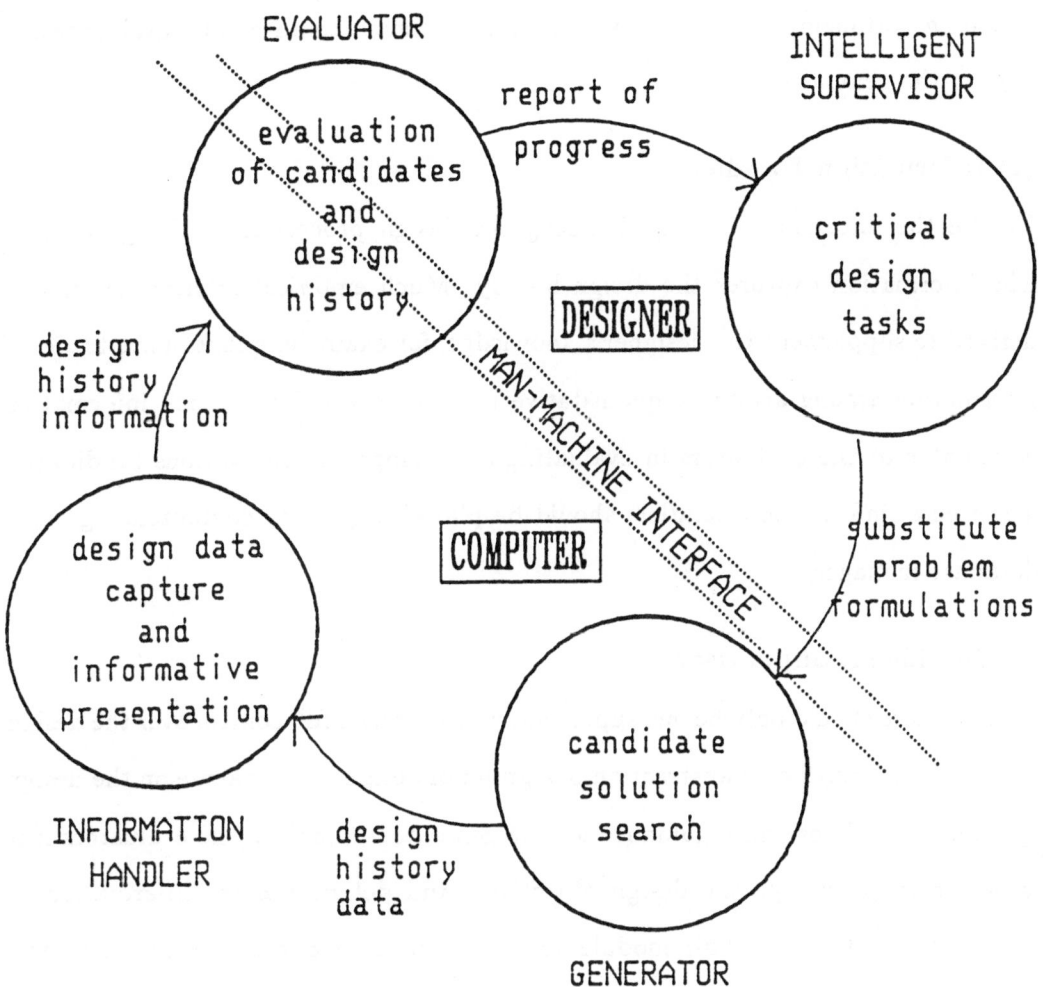

2.1 The Structure of GCS

inside an optimizing generator. We shall consider such an automatic evaluator as a part of the generator.

(c) Information Handlers

Handle the information spun off during the design process from the generators. The information captures the design history. Much empirical information can be derived to supplement the designer's knowledge, for example, quantitative trade-off information among arbitrary quantitative performance indices. Such information is valuable to the evaluators in evaluating and comparing the various candidates. The role of information handlers should be played largely by computers e.g. as a database manager.

(d) Intelligent Supervisor

Ideally, there should only be one supervisor who exercises full control over the design process. This may be an abstraction of a group of designers who agree on the design specifications. It provides guidance for the generators, modifies the evaluators if it is necessary to change the design objectives, and determines the information to be handled and saved. This module abstracts the most critical design tasks, viz. guiding the search, modifying design specifications to exercise trade-offs, altering the design approach, initiating and terminating the design process.

2.5 Man-machine Interface

Having established a structure for the GCS, we shall now consider man-machine interface, i.e. an appropriate sharing of tasks between the designer and the computer.

The intelligent supervisor naturally finds its place on the designer's side, and the information handlers on the computer's side. The evaluators span the interface since evaluation is a task best done co-operatively. The position of the generators is most flexible. For rapid generation, they should ideally be on the computer's side. However, there are many design methods which draw extensively on the designer's knowledge and analytic ability *during* the generation of any one design. For such methods, the generators would sit across the interface. The designer may also have designs generated entirely by himself. In this case, the generator is totally on his side.

2.6 Discussion

The development of the conceptual view of CACSD as GCS is now complete. We believe that a designer taking this conceptual view of CACSD will have a proper understanding of the design process. A design problem is naturally seen as composed of four sub-problems. This decomposition keeps a balanced view of CACSD and the designer will be more conscious of the role-play between him and the computer, Having taken up the role of an intelligent supervisor (and therefore the associated sub-problem of decision-making), the designer proceeds to solve CACSD problem by constructing the other modules between himself and the computer to tackle the other sub-problems. All the sub-problems are then co-ordinated and conducted as a GCS, viz. invoking proper generators to generate candidate designs of different subsets of the universe of controllers, registering them with the information handlers, evaluating them with the evaluators, updating his own knowledge about the problem and recursively guiding the generators to search for better designs.

2.7 Computing Supports for CACSD

Seen in the perspective of GCS, conventional packages on CAD are mostly concerned with implementations of design methods, i.e. those parts of their framework which correspond with the generators, while those with the other modules often receive little attention. A designer using such packages is therefore led to adopt conceptual views which are deficient, with the other modules almost out of sight. This may seriously impair the design as proper attention is not directed to the full scale operation of all the modules. Even if the designer has adopted our conceptual view, he will find such packages difficult to use since he will find minimal supports to the other three sub-problems. They are also inflexible since they may not yield to his idea of an efficient sharing of design tasks.

The goal of computing supports for the CACSD can be stated as the provision of a framework to correspond well with the GCS. Such a framework will provide the designer with a holistic supports of all the modules he requires in solving and co-ordinating the sub-problems.

2.8 IMOP as a Framework

In this work, we propose interactive multi-objective programming (IMOP) as such a suitable formal framework. First and foremost, it naturally handles the multi-objective nature of practical design. This is important since in practice, design problems are largely due to the need for trade-offs among conflicting objectives. Secondly, the various assumptions behind its development are relevant to practical design. For instance, its own development as a discipline in operational research is for the purpose of decision-making when there is much uncertainty in the user's objectives, and this is precisely the case with practical design. Thirdly, IMOP employs iterative optimizations driven interactively by the user in an explicit search for a solution. The concept of co-operative search is apparent. This should result in a natural correspondence with the conceptual view of the designer.

CHAPTER 3

INTERACTIVE MULTI-OBJECTIVE PROGRAMMING

3.1 Introduction

In this chapter, we shall elaborate on interactive multi-objective programming (IMOP) which we employ as the formal framework for the computer.

Interactive multi-objective programming (IMOP) is concerned with the class of multi-objective decision problems in which *"no single objective function can adequately serve to compare the difference in desirability among feasible solutions"* [Ros 85], and therefore a vector of performance indices has to be used to represent the multiple objectives. The decision problem is essentially a trade-off problem. IMOP is most appropriate when uncertainty abounds and when the set of feasible solutions is large, e.g. a compact non-empty subset of $\Re^n$. Man-machine interaction in the context of computer-aided decision making is often the only viable approach in such cases.

It has been noted that IMOP has many features similar to a man-machine interface [Saw 85]. The user is a decision-maker (DM). The computer is his analyst (AN) who conducts computations. An IMOP method is a scheme for communication between the DM and the AN by which the decision problem is progressively resolved by searching through the feasible alternatives and deciding on trade-offs among the objectives. They are sometimes called methods of progressive articulation of preference [Goi 82].

3.2 Formulation of IMOP

Formally stated, the basic decision problem is

$$\max_{\mathbf{x} \in X} \quad U\left(f_1(\mathbf{x}), f_2(\mathbf{x}), \ldots, f_q(\mathbf{x})\right) \tag{3.2.1}$$

where X is the set of feasible decisions, $f_1, f_2, \ldots, f_q$ are real-valued performance indices representing the multiple objectives and U is the utility function which represents the preference structure of the decision-maker (DM). We also define the index set

$$Q = \{1, 2, \ldots, q\} \tag{3.2.2}$$

In the performance index space, the decision problem may be stated as

$$\max_{\mathbf{y} \in Y} \quad U(\mathbf{y}) \tag{3.2.3}$$

where $Y = \{(y_1, \ldots, y_q)\} = \{(f_1(\mathbf{x}), \ldots, f_q(\mathbf{x})) : \mathbf{x} \in X\}$ is the set of feasible performance index vectors.

The utility function is a conceptual construct which represents the degree of the DM's satisfaction with the attainment of a particular vector of performance indices. An analytic expression may not exist. Without loss of generality, we assume U to be strictly increasing with decreasing values of any of the performance indices. Without being mathematically strict, we can again restate the problem as a vector minimization problem

$$\min_{\mathbf{y} \in Y} \quad \mathbf{y} \tag{3.2.4}$$

The need for interaction is due to the fact that the information required for solving the problem is (1) decentralized in the sense that the DM knows about U and the AN knows about Y and (2) not immediately available in the sense that explicit expressions of U and Y do not exist and no global information about either U or Y is available to both DM and AN. Therefore, the AN will produce computed elements of the feasible set as possibilities on the basis of which the DM will articulate his wishes.

3.3 Solution Concepts (Fig. 3.1)

An important concept of multi-objective programming is solution efficiency. A solution to the vector minimization problem (3.2.4), x^* is efficient if and only if

$$\forall x \in X, \qquad f(x^*) - f(x) \in \Re^q_+ \quad \Rightarrow \quad f(x^*) = f(x) \tag{3.3.1}$$

where $\Re^q_+$ is the non-negative orthant of $\Re^q$. (y_1 is said to dominate y_2 iff $y_2 - y_1 \in \Re^q_+$.) Efficient solutions, in general, are not unique. The set of efficient solutions is called the efficient set, denoted by X^*. (Other equivalent terms are in use, e.g. non-dominant solutions, Pareto-optimal solutions, non-inferior solutions, etc). For q performance indices, the efficient set will project into hypersurfaces in the performance index space of at most $(q - 1)$ dimensions. The union of these hypersurfaces, which is the range of X^* in Y, is called the efficient frontier, denoted by Y^*. An example of an efficient frontier for the two index case is illustrated in fig. 3.1. In general, the larger is Y^*, the greater is the need for interaction between the DM and the AN to seek a good compromise solution among the efficient solutions.

The problem (3.2.4) only defines X^* but is not itself a computable formulation. A scalarization is needed for generating efficient solutions using auxiliary optimizations. The general form of a scalarization is

$$\min_{x \in X} \quad F(y, \alpha) \tag{3.3.2}$$

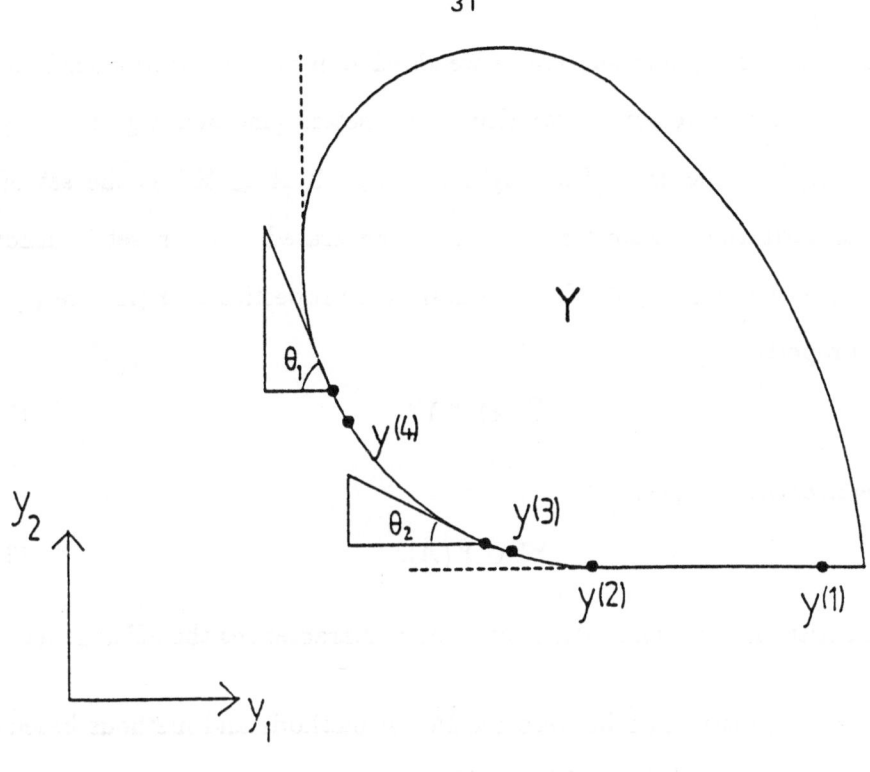

$$y^{(1)} \in \tilde{Y} - Y_*$$

$$y^{(2)}, y^{(3)}, y^{(4)} \in Y^{\bullet}$$

$y^{(4)}$ is proper efficient, if so defined as the marginal rate of substitution (MRS) be bounded by $\tan \theta_2 \geq MRS \geq \tan \theta_1$

3.1 Solution Concepts of Multi-Objective Programming : Two-Index Case

where F is a scalarizing function (e.g., a weighted sum of the performance indices, $\sum_{i=1}^{q} w_i f_i$) and α is a vector of scalarizing parameters (the non-negative weights $(w_1, w_2, \ldots, w_q)^T$ in the case of a weighted sum). If $A \subset \Re^m$ is the set of all admissible scalarization parameter vectors, the generated solution set is denoted as $\tilde{Y}(A) = \{y(\alpha) : \alpha \in A\} \subset Y$. Two desirable properties of $\tilde{Y}(A)$ are (i) the sufficiency property

$$\tilde{Y}(A) \subset Y^* \tag{3.3.3}$$

and (ii) the necessity property

$$Y^* \subset \tilde{Y}(A). \tag{3.3.4}$$

If both are satisfied, the scalarization completely characterizes the efficient set.

An important distinction between the IMOP methods and methods based on the multi-attribute utility theory [Nak 85] lies in their use of the scalarizing functions. For the latter approach, the primary aim is to approximate the utility function U with F. For the IMOP methods, the scalarizing functions are merely means to generate efficient solutions based on the DM's currently expressed wishes.

A weaker condition than (3.3.1) is weak efficiency when $\Re_+^q$ is replaced by $\dot{\Re}_+^q \cup \{0\}$, where $\dot{\Re}_+^q$ is the positive orthant of $\Re^q$. Nearly all, if not all, weakly efficient solutions are efficient. If the weakly efficient frontier is $\tilde{Y}^*$, the exceptions are

$$\left\{ x : \exists \tilde{y} \in \tilde{Y}^*, \quad f_k(x) > \tilde{y}_k, \quad f_i(x) = \tilde{y}_i, \quad k \in Q_k \subset Q, \quad \forall i \in Q/\{Q_k\} \right\} \tag{3.3.5}$$

which corresponds to the presence of "straight edges" in the efficient frontier which are parallel to some of the co-ordinate axes (fig. 3.1). These exceptions are rare when the f_i's are nonlinear maps.

A stronger condition than (3.3.1) is proper efficiency which rules out efficient solutions with too large or too small marginal rates of substitutions between the indices (fig 3.1). However, definitions are many and often subjective, depending on the assigned bounds for the marginal rates.

3.4 The Basic Algorithm of IMOP Methods

IMOP methods are interactive algorithms. Unlike conventional algorithms which are effective procedures for automatic computations, interactive algorithms comprise intermediate steps by which the user affects the course of computation. The structure of IMOP's basic algorithm is a three-step interactive algorithm:

STEP 0 : Initialize

A scalarization of (3.2.4) is set up for the AN as its means to generate efficient solutions. AN computes initial information about possibilities. DM supplies initial information about his wishes;

STEP 1 : AN-Generate

The AN generates one or more efficient solutions from the scalarization and presents resulting information to the DM;

STEP 2 : DM-Trade-off

The DM expresses his wishes by soliciting trade-off information concerning the generated solutions, resulting in a change of the scalarization parameters.

The last two steps are repeated until the DM indicates acceptability of a current achievement level, provided one exists.

In general, IMOP methods vary in the scalarizations they use, schemes for communication between DM and AN, and the trade-off analysis methods.

3.5 IMOP and CACSD

We have noted before that IMOP is most suitable for multi-objective decision problems when uncertainty abounds. Goicoechea [Goi 82] elaborated this by pointing out that IMOP is predicated on the following assumptions about the psychology of the decision-making process:

(1) Gestalt Philosophy

DM's perception is influenced by the partial set of solutions he has encountered so far. This is an assumption of Gestalt philosophy.

(2) Implicit Utility

The utility function cannot be expressed analytically, although it is assumed that the DM does subscribe to a set of beliefs.

(3) Dynamic Preference Structure

The DM's preference structure changes over time as a result of learning and experience.

(4) Satisficing

The DM normally satisfices rather than optimizes and a solution to a decision problem is any acceptable course of action.

(5) Learning

Acceptability is a learned perception.

It is not difficult to see the relevance of these assumptions for CACSD.

Broadly speaking, practical design problems are multi-objective decision problems for which uncertainty is always an intrinsic feature. There exist only principles and guidelines (assumption (2)) and the designer, and his customers, are often learning about possibilities and impossibilites of the design problem (assumption (5)). Such learning is based on the analysis of candidate designs generated during the design process which results in a continual modification of specifications (assumptions (1) and (3)). The final specifications, which are criteria for acceptance of a candidate, are never fully determined *a priori* but are obtained as a result of learning (assumption (5)). The final design is one which satisfies the specifications to an acceptable degree. Optimization of objectives is often impractical due to design cost constraints (assumption (4)).

3.6 From The Conceptual View to A Formal Framework

Apart from the relevance of IMOP's assumptions for CACSD, we can also readily appreciate the similarity between the basic algorithm of IMOP and the structure of the generalized co-operative search. Most importantly, the basic algorithm features the information feedback which is the essential link between the various modules in the conceptual view (fig. 2.1). If the designer is engaged in an interaction with an explicit structure of the basic algorithm, he will find a good correspondence with his conceptual view of CACSD problems. The generators and the information handlers are active in the AN-Generate step while the evaluators and the intelligent supervisor are active in the DM-Guide step. The interaction of these two steps complete the path for information feedback.

Therefore we propose to use IMOP as the framework to organize various design facilities on the computer. The designer-computer interaction is carried out with a structure of IMOP's basic algorithm through which the facilities are accessed.

3.7 Components of IMOP Methods

Having established IMOP as a suitable framework of the computer for the purpose of CACSD, we shall now examine the components of IMOP methods critically.

We shall look at three components of the IMOP methods which are particularly relevant to our conceptual view of CACSD. They are scalarizations, communication schemes and trade-off analysis methods.

3.7.1 Scalarizations

In our conceptual view of CACSD, scalarizations constitute a class of generators to which the intelligent supervisor provides guidance in terms of the scalarization parameters. They are the instrument by which efficient solutions are generated in the AN-Generate step of the basic algorithm. In addition to (3.3.3) and (3.3.4), it is also a desirable feature for the scalarizing parameters to be meaningful to the DM, in the sense that (i) they have a natural interpretation and (ii) they give the DM *a priori* control of the outcomes.

The most commonly used scalarizations are :

(1) **Weighted Sums** (Fig. 3.2)

$$\min_{x \in X} \left\{ w^T f(x) : w = (w_1, w_2, \ldots, w_q)^T \in \Re_+^q ; \quad \sum_{i=1}^q w_i = 1 \right\}$$

$$= \min_{x \in X} F_w(f(x), w) \tag{3.7.1.1}$$

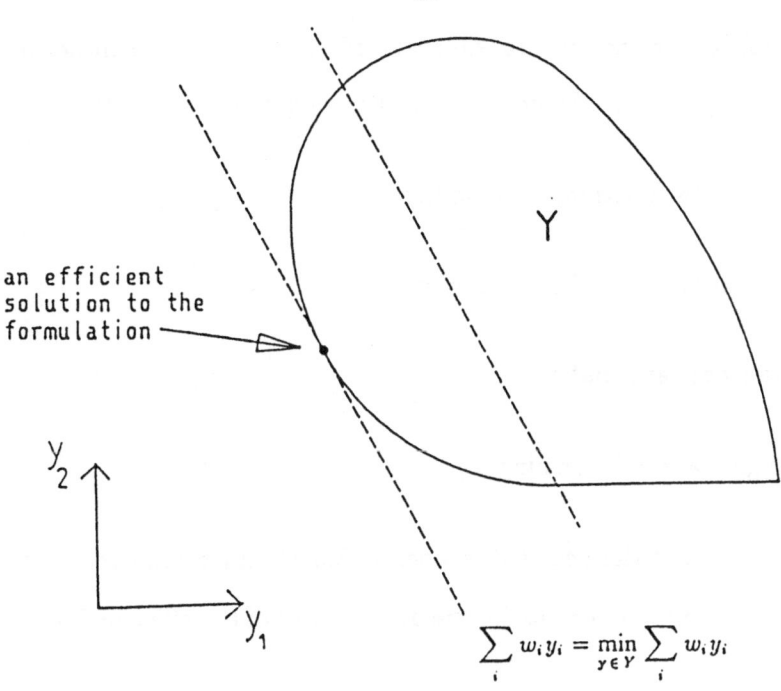

an efficient
solution to the
formulation

Y

y_2

y_1

$$\sum_i w_i y_i = \min_{y \in Y} \sum_i w_i y_i$$

3.2 Scalarization : Weighted Sum Formulation

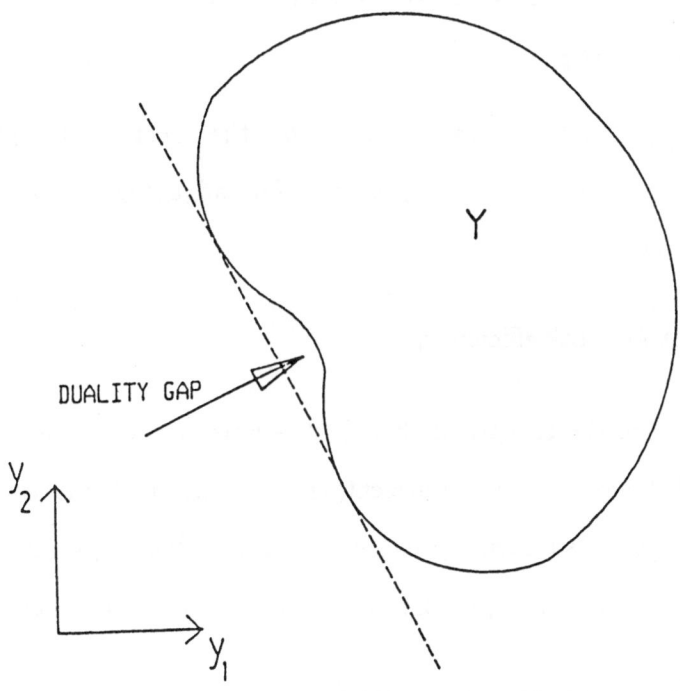

Y

DUALITY GAP

y_2

y_1

3.3 Duality Gap

The scalarizing parameters are elements of the vector of non-negative weights $w = (w_1, w_2, \ldots, w_q)^T$. The advantages of a weighted sum are that :

(i) it is a smooth function of the indices;

(ii) it is sufficient for efficiency when $w \in \dot{\Re}_+^q$.

The disadvantages are that :

(i) the weights are not meaningful;

(ii) it does not have the necessity property (3.3.4) and cannot reach efficient solutions in the concave regions (or the duality gap) of the efficient frontier (fig. 3.3).

(2) ϵ-Constraint Formulations (Fig. 3.4)

$$\min_{x \in X} \{f_k(x) : f_i(x) \le \epsilon_i, \quad k \in Q, \quad \forall i \in Q/\{k\}\}$$
$$= \min_{x \in X} F_\epsilon(f(x), k, \epsilon) \tag{3.7.1.2}$$

The scalarizing parameters are elements of the vector of index bounds $\epsilon = (\epsilon_1, \epsilon_2, \ldots, \epsilon_q)^T$ with ϵ_k a dummy entry. The advantages of an ϵ-constraint formulation are that :

(i) it is sufficient for weak efficiency;

(ii) in the absence of the exceptions (3.3.5), it is both necessary and sufficient for efficiency. If the exceptions are present, the necessity and sufficiency property can be restored by successively executing the auxiliary optimizations of the formulation for $k = 1, 2 \ldots, q$, while setting $\epsilon_k = \min_{x \in X} F_\epsilon(f(x), k, \epsilon)$;

(iii) the scalarizing parameters are meaningful.

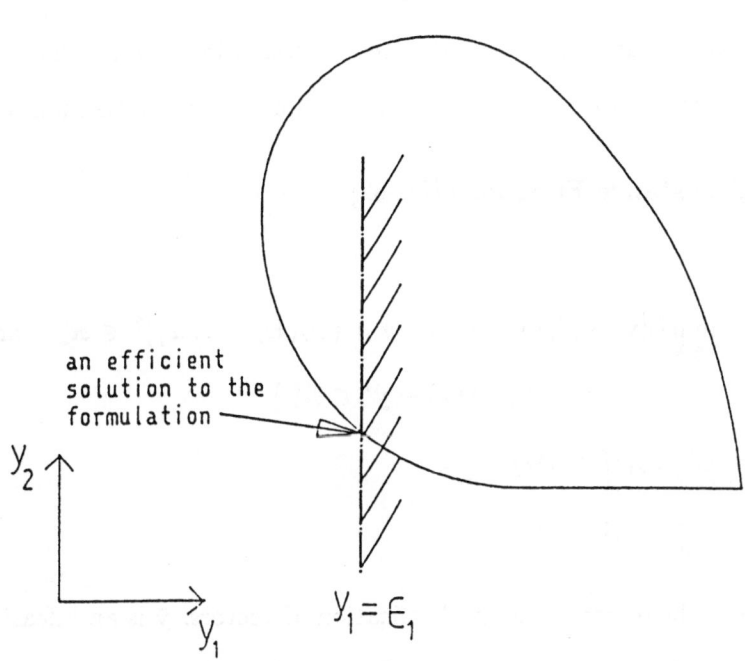

an efficient
solution to the
formulation

$y_1 = \epsilon_1$

3.4 Scalarization : ϵ-Constraint Formulation

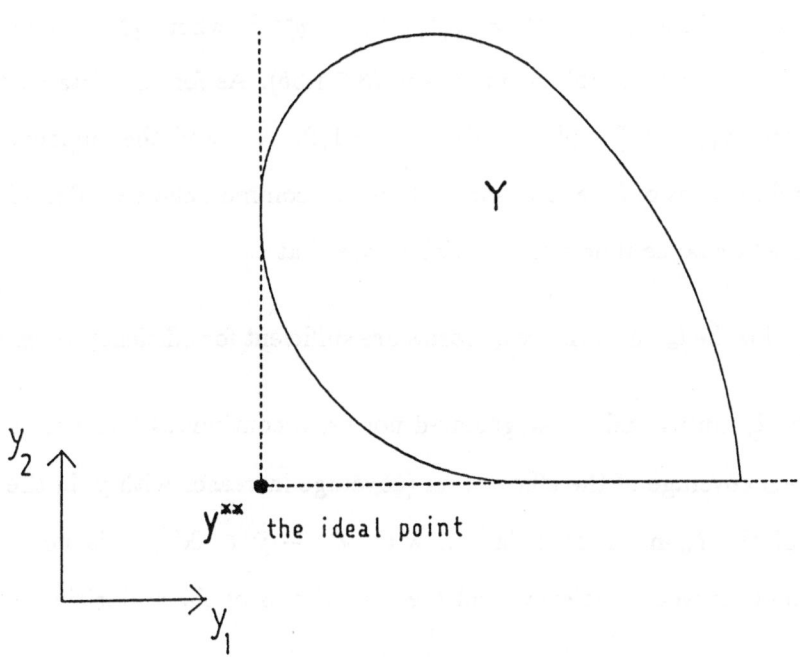

Y

y^{**} the ideal point

3.5 The Ideal Point

The disadvantage is that being a constrained optimization, the auxiliary optimizations are more complicated in terms of computation and run-time maintenance.

(3) Weighted Distance Functions [Zel 82]

$$\min_{x \in X} \{ \| \mathbf{w}. * (\mathbf{f}(x) - \mathbf{y}^*) \| : \mathbf{w} = (w_1, w_2, \ldots, w_q)^T \in \Re^q_+ \quad \text{and}$$

$$\forall x \in X, \quad \mathbf{f}(x) - \mathbf{y}^* \in \Re^q_+ \}$$

$$= \min_{x \in X} F_{d_1}(\mathbf{f}(x), \mathbf{w}) \tag{3.7.1.3a}$$

$$\text{or} \quad = \min_{x \in X} F_{d_2}(\mathbf{f}(x), \bar{\mathbf{y}}) \tag{3.7.1.3b}$$

where .* denotes the element-wise multiplication of vectors, $\bar{\mathbf{y}}$ is an infeasible point and $\mathbf{w}$ is a vector of non-negative weights. The weighted distance function measures the deviation from $\bar{\mathbf{y}}$. The scalarizing parameters are either (i) elements of the vector of non-negative weights $\mathbf{w} = (w_1, w_2, \ldots, w_q)^T$ as in (3.7.1.3a), in which case $\bar{\mathbf{y}}$ is often set to the ideal point $\mathbf{y}^{**} = (y_1^{**}, y_2^{**}, \ldots, y_q^{**})^T$ where $y_i^{**} = \min_{x \in X} f_i(x)$ (fig. 3.5), or (ii) the infeasible point $\bar{\mathbf{y}}$ as in (3.7.1.3b). As for the distance function, the l_p-norms $\| \mathbf{y} \|_p = (y_1^p + y_2^p + \cdots)^{1/p}$, $p = 1, 2, \ldots$ and the augmented norms (linear combinations of l_1- and l_∞-norms) are the common choices. The advantages of a weighted distance function formulation are that :

(i) except for the l_∞-norms, the l_p-norms are sufficient for efficiency when $\mathbf{w} \in \dot{\Re}^q_+$;

(ii) for the l_p-norms and the augmented norms, a continuous trade-off is possible between coverage of the efficient set (coverage increases with p; in the limiting case of the l_∞-norm formulation, with $\mathbf{y}^{**} - \bar{\mathbf{y}} \in \Re^q_+$, it is necessary and sufficient for weak efficiency) and the smoothness of the scalarizing function;

(iii) the scalarizing parameters are meaningful, representing either the relative importance of the attainment of the infeasible point's components in (3.7.1.3a) or a utopian goal of the DM (3.7.1.3b).

The disadvantages are that :

(i) initial parameters which affect solution outcomes have to be determined *a priori* ($\bar{y}$ in the case of (3.7.1.3a) and w in the case of (3.7.1.3b));

(ii) efficient index vectors not dominated by $\bar{y}$ cannot be generated.

(4) Achievement Functions [Wie 82]

$$\min_{x \in X} F_a(f(x) - \bar{y}) \qquad (3.7.1.4)$$

where $\bar{y}$ is a reference point in the index space. Unlike the infeasible reference point of the weighted distance function, $\bar{y}$ does not have to be infeasible. An achievement function serves to penalize deviations from $\bar{y}$, except in cases of over-attainment, when it allocates the surplus. Achievement functions are continuous. We restrict our attention to a class of popular achievement functions which are (i) weakly monotonic, i.e.,

$$\forall y_1, y_2 \in \Re^q, \quad y_2 - y_1 \in \dot{\Re}^q_+ \quad \Rightarrow \quad (\forall \bar{y} \in \Re^q, \quad F_a(y_1 - \bar{y}) < F_a(y_2 - \bar{y}))$$

$$(3.7.1.4a)$$

and (ii) order-representing, i.e.,

$$\bar{y} - f(x) \in \Re^q_+ \quad \Rightarrow \quad F_a(f(x) - \bar{y}) \leq F_a(0). \qquad (3.7.1.4b)$$

A common class of weakly monotonic order-representing achievement functions is

$$F_a(f(x) - \bar{y}) = \begin{cases} F_{d2}((f(x) - \bar{y})_+, 0) & \text{if } f(x) - \bar{y} \in \Re^q_+ ; \\ F_m(f(x) - \bar{y}) & \text{otherwise.} \end{cases} \qquad (3.7.1.4c)$$

where the elements of $(f(x) - \bar{y})_+$ are $\max((f_i(x) - \bar{y}_i), 0)$, $i = 1, 2, \ldots, q$ and $F_m : \Re^q \rightarrow R$ is a strongly monotonic function (condition (3.7.1.4a) with $\Re^q$ replaced by $\Re^q_+ / \{0\}$), e.g. $-\prod_{i=1}^q (\bar{f} - f(x))$ (fig. 3.6). The advantages of this class of achievement functions are that

(i) they are necessary and sufficient for weak efficiency;

(ii) they are necessary for efficiency, with every efficient solution x^* solving $\min_{x \in X} F_a(f(x) - f(x^*))$;

(iii) the parameters are meaningful, representing the goal of the DM: (a) in the case of infeasible $\bar{y}_i$, a weakly efficient solution will be obtained which is nearest to it (in the sense of the weighted distance function F_{d2} in the subspace of the infeasible components of $\bar{y}_i$); (b) when efficient, it will be attained exactly; and (c) when over-attainable, an efficient solution which dominates it will result;

(iv) the optimal value of the achievement function, F_a^* for a set reference point $\bar{y}$, is also meaningful: $F_a^* > 0$, $F_a^* = 0$ and $F_a^* < 0$ implies $\bar{y}$ infeasible, efficient and over-attainable respectively.

The disadvantage is that the function is always non-smooth at $f(x) = \bar{y}$ due to the axioms of strong-monotonicity and order-representing [Wie 85].

3.7.2 Communication Schemes

According to Bogetoft [Bog 86], *"... the (IMOP) procedures point out the essential judgments to make and the necessary information to submit to make the communication progress. The focal point is to balance approximate substitution wishes agaianst approximate substitution possibilities."*

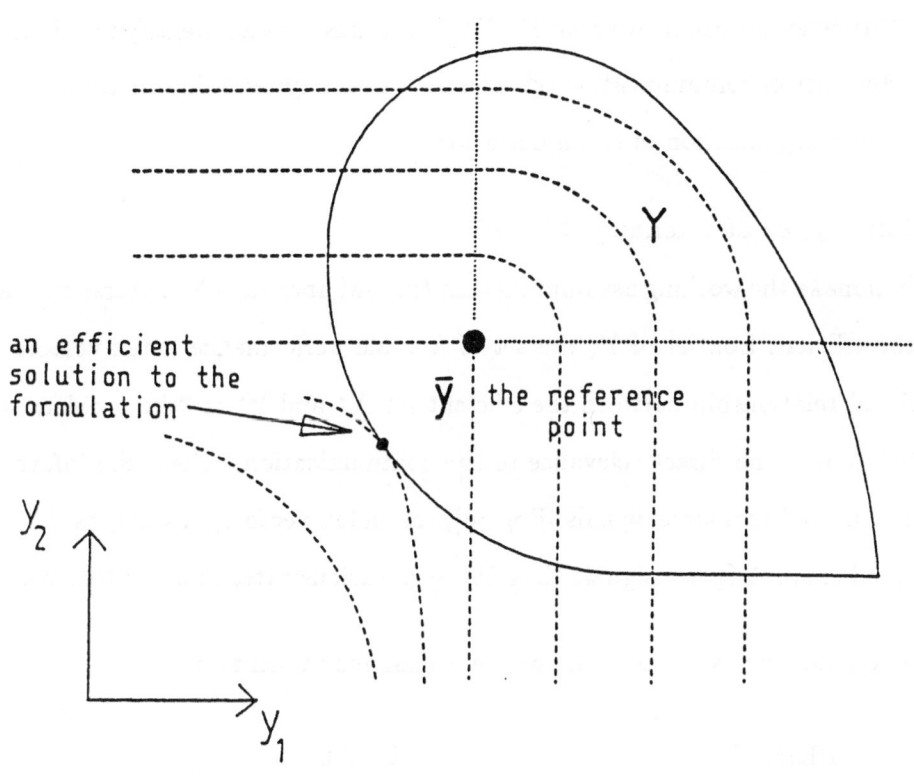

3.6 Scalarization : Achievement Function Formulation

From this general statement about IMOP procedures, we can identify two distinctive characteristics of communication schemes as: (1) the types of information exchange and (2) the organization of communication.

(1) The Types of Exchanged Information

We shall make the working assumption that the DM and the AN converse only about Y^*, the efficient frontier of Y, the set of feasible performance index vectors. The functional relationship between the efficient set X^* and Y^* is "the problem within the AN" and of no direct relevance to the communication. Then, the information is either primal (resource signals [Bog 86]), as index vectors, or dual, as directions in the index space (price signals [Bog 86], e.g. various rates of substitution).

The following table summarizes the exchanged information:

	PRIMAL	DUAL
DM	specifies target regions or points in the index space for attainment;	specifies substitution wishes, e.g. "1 unit of f_1 for every two units of f_2";
AN	presents attained index vector and/or regions around them;	presents possible rates of substitutions, i.e. orientation of supporting hyperplanes at the efficient frontier;

The information that the DM solicits is strongly influenced by the scalarization. For weighted sums and weighted distance functions (3.7.1.3b), substitution wishes are often used (e.g. Ziont-Wallenius method [Zio 83], pairwise comparison methods [Kok 85])). For ϵ-constraint formulations, weighted distance functions (3.7.1.3a) and achievement functions, targets are often used (e.g. interactive multiple goal programming [Spr 81], ideal displacement method [Zel 82], reference point method of Wierzbicki [Wie 82]). However, there are exceptions, e.g. the interactive satisficing method employs (3.7.1.3b) with l_∞-norm while requesting target points from the DM.

(2) The Organization of Communication

Communication is needed because the information required for solving the decision problem (3.2.3) is distributed and not readily available. The DM and the AN have to communicate about possibilities (Y) and wishes (U) and the communication shall be organized for

(1) the DM to learn about Y (DM-Directed [Bog 86]),

(2) the AN to learn about U (AN-Directed [Bog 86]), and/or

(3) the DM and the AN to reach a consensus.

Such organization appears as an extra structure imposed on the basic algorithm in section 3.4.

To be DM-directed, *"the AN should provide an interactive instrument with which the DM can probe, sample, and wander through the efficient set at his or her own discretion"* [Ros 85]. We also propose that he be supported in visualizing the efficient frontier to have a more holistic view of the limit of achievement. This is an important argument we put forward and design tools introduced in chapter 7 are based on this.

To be strictly AN-directed is not too desirable when the assumptions of implicit utility and dynamic preference structure in section 3.5 are made. However, it should be possible, and useful, to support the DM-directedness by helping to identify undesirable regions in the efficient frontier, and with the DM's approval, narrow the search down to the desirable regions.

To reach a consensus, they have to go through iterations of AN-Generate and DM-Trade-off (in the basic algorithm of section 3.4) until the DM is satisfied with his choice. However, there is *"a trade-off between the complexity of interaction in each iteration and the number of iterations"* [Kok 86]. A basic observation in psychology is that human information processing capabilities are limited. Experiments have shown that as the amount of information given to the DM increases, the percentage of information used decreases [Pay 76]. The important issue here is how to strike the balance by assisting the DM-Trade-off step with an appropriate amounts of effective information.

3.7.3 Trade-off Analysis Methods

Trade-off analysis methods utilize certain types of information and display them in textual or graphical forms composing the ground on which the DM decides on his trade-off wishes. The methods help the DM to evaluate the efficient solutions, learn about possibilities, and articulate his wishes more effectively.

A basic observation of cognitive psychology is that humans have excellent pattern recognition capabilities and are always good at digesting visual data. Therefore, graphical displays of data complemented by textual information are accepted to be superior over purely textual display [Tuf 83]. In the trivial case of two performance indices when the efficient frontier is a curve in the index space, the best method is for the AN to display the curve graphically and the DM to make his choice among points in the visible frontier. Only the primal information is presented and *the dual information can be readily perceived* (fig. 3.7).

Although there are successful attempts at high dimensional data display [Tuf 83], interpreting them can be a highly involving exercise for the user. For this reason, dual information is often explicitly presented for trade-off, with textual display as the major, if not the only, medium. Various forms of dual information have been suggested, e.g. marginal rates of substitution, exaggerated rates of substitution, Kuhn-Tucker multipliers etc.

However, there are two disadvantages with dual information. Firstly, the substitution possibilities they carry are only local, when the efficient frontier is non-linear, and often exaggerating, when the efficient frontier is convex. Secondly, they are cognitively more distant from the DM than the primal information are. It is the result of an experiment concerning the use of various IMOP methods that although dual information was made available, the DMs did not pay much attention to this [Kok 85].

There are also methods analogous to that of the two index case mentioned in which a subset of index vectors on the efficient frontier are displayed. Broadly speaking, there are two alternatives to do so. The components of the vectors are plotted against either (i) a permutation of elements of the index set Q or (ii) an ordering of the vectors (fig. 3.8).

Apart from trade-off analysis methods which originate from IMOP methods, there also are independent developments from the area of multi-objective (multi-criteria) analysis, which aims to assist an explicitly multi-objective evaluation of a finite number of feasible alternatives [Roy 81]. These include the outranking meth-

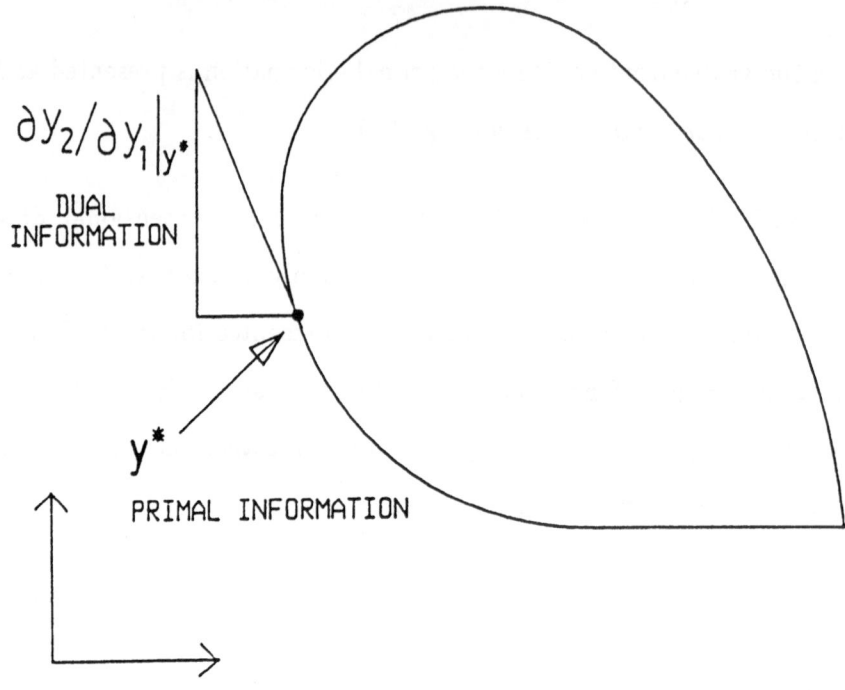

$$\partial y_2 / \partial y_1 \big|_{y^*}$$

DUAL
INFORMATION

y^*

PRIMAL INFORMATION

3.7 Primal and Dual Information : Two Index Case

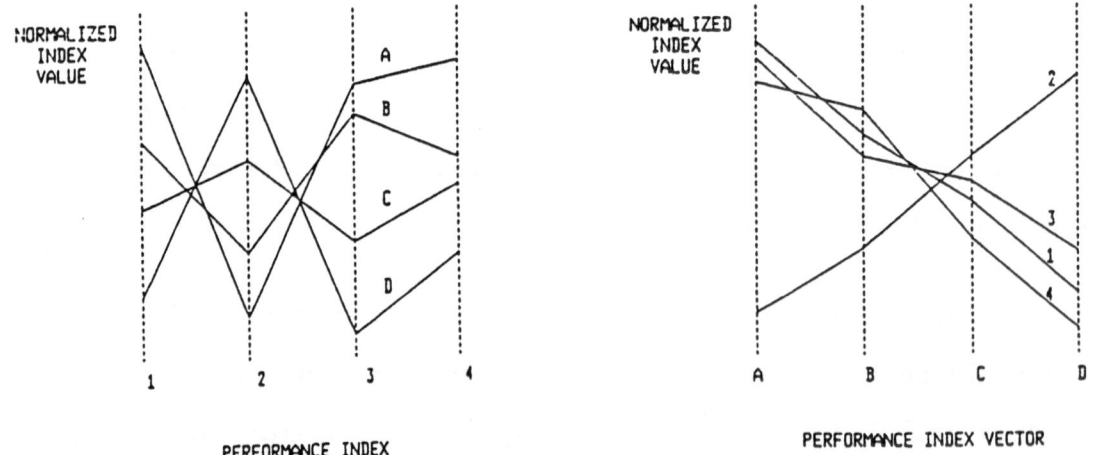

3.8 Graphical Presentations of Efficient Index Vectors

ods [Bra 84], entropy-based measures of objective weights [Nak 85], choice function approach based on fuzzy set theory [Orl 78], etc. These methods invariably aim to deduce an ordering among either the objectives or the alternatives, establishing a scale of importance among the objectives, or identifying dominated solutions. They are not devised for providing guidance to the further generation of efficient solutions. Even if they do, the resulting interaction may tend to be too AN-directed.

3.8 Discussion

It is not difficult to appreciate the potential of the various concepts and techniques of IMOP in their application to CACSD as suggested in sections 3.5 and 3.6. In chapter 5, we shall develop a strategy for CACSD when the basic algorithm of IMOP is used as a framework to organize the design facilities. The collection of concepts and techniques as surveyed in this chapter and the next will be selected for use in the strategy.

CHAPTER 4

SEARCH-BASED CONTROL SYSTEM DESIGN METHODS

In this chapter, we survey the class of search-based methods for control system design. This class of methods execute design by searching in controller spaces with Euclidean structures using numerical search techniques, and in particular, the iterative optimization algorithms. As mentioned in chapter 2, this is the class of methods which best supports the design process as a whole since the concept of search, which they are based upon, is "the most useful single characterisation". We shall evaluate how well these methods fit into the conceptual view, in particular, we shall see how effective they are as generators.

We have identified four distinguished approaches within this class. We shall describe them in such a way as to highlight their respective emphasis and then evaluate them as generators in the designer's view.

4.1 Zakian's Method of Inequalities [Zak 73]

Zakian and Al-Naib first proposed the formulation of a controller design problem as a general set of inequalities, i.e. find x such that

$$f_i(x) \leq c_i \qquad i = 1, 2, \ldots, q \qquad x \in \Re^n \qquad (4.1.1)$$

where x is a vector of design parameters parametrizing a controller with a particular structure. The f_i's are the performance indices which represent design objectives. Some of the f_i's are majorants which serve to replace functional constraints which are constraints of the form

$$g_i(x, z) \leq 0 \qquad \forall z \in Z_i$$

where W_i is a compact set, e.g. a segment in the real line. In such cases, if

$$f_i(\mathbf{x}) \geq g_i(\mathbf{x}, z) + c_i \qquad \forall \mathbf{x} \in \Re^n, z \in Z_i \qquad (4.1.2)$$

we say $f_i(\mathbf{x})$ majorizes g_i.

Functional constraints are not uncommon in control system design, e.g. template bounds for frequency responses and time responses. Development of such majorants has been an emphasis of Zakian's work in this area [Zak 84].

A simple numerical search algorithm called the moving boundary process is used to obtain the numerical solution of the inequalities. In cases when the search cannot locate such a solution, it will try to satisfy as many of the inequalities as possible. There have been reports on successful applications of this method [Tai 86], [Bad 87]. A controller structure is selected first. Simple ones are preferred but may be progressively augmented in complexity. The designer's role includes prescribing the bounds c_i's for the indices and changing them if necessary. Its application to design thus proceeds as a sequence of such formulations.

It is also emphasized that the indices should represent natural measures of performance. (An important index that Zakian proposed for linear systems is the least upper bound of an error response over all time and all possible inputs. It turns out that if D is the supremum of the absolute rate of change of all possible inputs, the index becomes simply $D * IAE$ where IAE is the integral absolute error of the response to a step).

Evaluations

The strength of the approach is its simplicity: a small set of inequalites carefully chosen with natural measures of performance and a high-level "black-box" search algorithm. The interface between the generator and the intelligent supervisor is simple. Driving the generator involves changing the bounds of the set of inequalities, which is a simple task in itself.

Yet the set of inequalities on its own can be an inappropriate simplification when the attainment of all the assigned bounds is stringent and even impossible. In such cases, there is a need to establish some measure of the relative importance of the bounds to affect the order of their attainment by the moving boundary process. However, with the formulation (4.1.1), the only way to reflect the relative importance is by co-ordinating the "relative tightness" of the bounds which depends not only on the bounds' values but also on the scaling of the indices. Even if this tightness is co-ordinated, the order of attainment still depends on the relative ease for the search algorithm to improve the index values. In actual fact, what is needed in such a case is to establish a metric in the index space by a scalarization (e.g. the weighted distance function (3.7.1.3)).

To avoid the above mentioned problem, the method of inequalities is often initiated with a set of loose bounds. In the subsequent formulations, the bounds are progressively tightened up while making sure that there are solutions to the inequalities.

4.2 Semi-Infinite Optimization

This approach due to Polak, Mayne and co-workers aims at developing special optimization algorithms. For a certain class of problems, the convergence of these algorithms can be proven, to deal with the functional constraints (4.1.2) which

Zakian bypassed using majorants. The canonical problem which they tackle is

$$\min_{x \in \mathbb{R}^n} f_0(x)$$

$$\text{subject to} \qquad f_i(\mathbf{x}) \leq 0 \qquad i = 1, 2, \ldots, p$$

$$\text{and} \qquad g_j(\mathbf{x}, z) \leq 0 \qquad j = 1, 2, \ldots, r \tag{4.2.1}$$

$$\forall z \in Z_j \subset R$$

where $g_j(x, z)$ are the functional constraints.

Apart from algorithms for solving such semi-infinite optimization problems, the Polak-Wardi [Pol 82] and Polak-Stimler algorithms [Pol 84] were specifically conceived for control system design. A set of functional performance indices were derived [Pol 84]. For instance, input tracking is specified as

$$b^l(t) \leq y(x, t) \leq b^u(t) \qquad \forall t \in (0, T_s) \tag{4.2.2}$$

which is a requirement for the time trajectory $y(x, t)$ (e.g. a step response) to be within a prescribed region defined by $b^l(t)$ and $b^u(t)$ over the transient period $(0, T_s)$ (fig. 4.1). S-stability (placing all poles of a linear system within prescribed region S in the complex plane, fig. 4.2), if S has linear edges, can be specified as

$$Re\left(e^j(A(\mathbf{x}))\right) + b * Im\left(e^j(A(\mathbf{x}))\right) + c \leq 0 \qquad \text{for} \quad j = 1, 2, \ldots, N_c \tag{4.2.3}$$

where the e^j's are eigenvalues of $A(x)$.

A set of majorization techniques are also developed in conjunction by Polak and Stimler [Pol 86]. Unlike Zakian's majorants, these majorants are still functionals, but which are computationally simpler than the original ones. For instance, the frequency response of an uncertain plant at selected frequency points can be

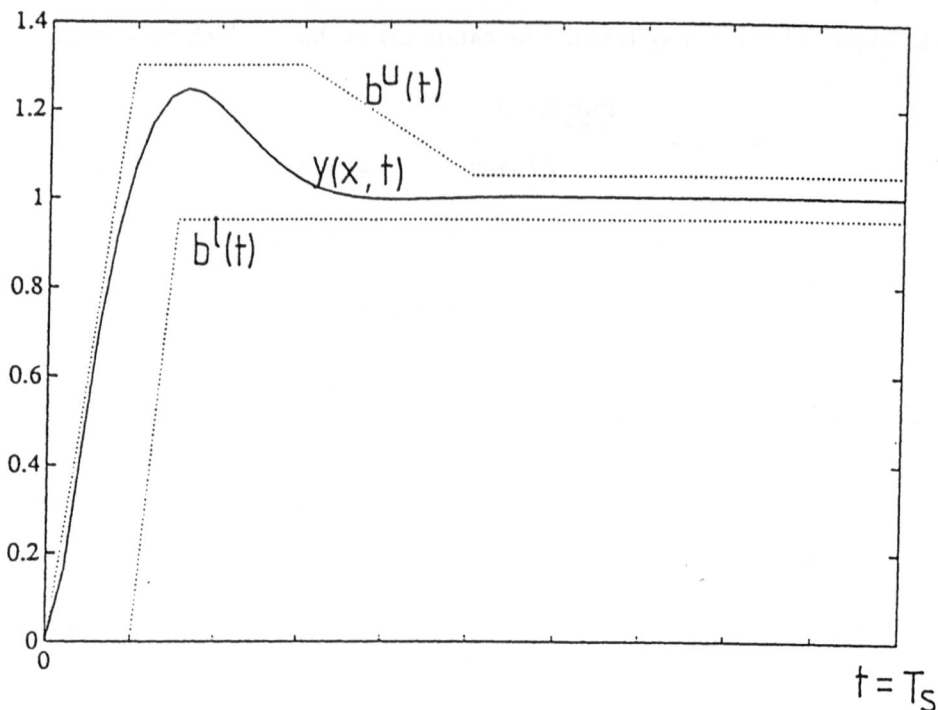

4.1 Functional Constraints for a Step Response

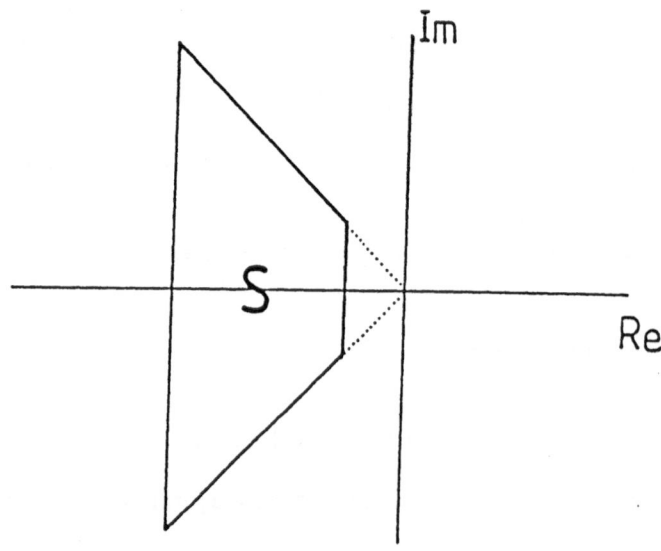

4.2 Functional Constraints for Close Loop Poles : S-Stability

represented as a set of rectangles which majorize the original set composed of closed regions, whose boundaries are curved in general.

Evaluations

Semi-infinite optimization expands the formulation of classical optimization at the expense of execution and maintenance costs due to the functional constraints, while the gain is in the precision of the satisfaction of them. In most practical design cases when the formulations are not accurate, prescribing templates for functional constraints may involve too much effort in itself, not to mention their modification when trade-offs are required. Trade-offs for functional constraints are often expected since very often, most, if not all, of them are not hard constraints. For instance, seldom does a problem requires a loop transfer function to be strictly within a template; slight violation is often acceptable.

Much has to be done to reduce the interface complexity (e.g. the assignment of functional bounds by graphical input) and enhance the robustness of these algorithms before they can be applied to the practical design environment. A simpler optimization formulation which replaces the functional constraints by a set of discretized ordinary ones is often found to be adequate for exploring the design problem as well as generating controllers of reasonable performance [Ng 87]. In most practical designs, especially when the design problem is being explored for trade-offs, the degree of precision that semi-infinite optimization offers may simply be inappropriate. The simplicity of the optimization algorithm is a more important consideration.

However, it should be appreciated that such generators will be most suitable for the last stage of design when uncertainties have been clarified, most trade-off decisions made, the attainability of the specifications better understood, and a controller structure decided with a good initial set of parameters. Only when an all-embracing and accurate formulation is available will such algorithms, which guarantee convergence, be suitable.

4.3. Multi-Objective Programming

We identify methods of this approach as those which seek to formulate control system design as explicit multi-objective programming problems. Their works are concentrated on the scalarization with minimal attention to the other two aspects we cited in the last chapter, viz. the communication schemes and trade-off analysis methods.

Tabak et al. implemented a trade-off algorithm for generating an efficient solution based on a set of previously obtained ones [Tab 79]. They are also trying to solve a set of inequalities as Zakian does. They differ however in that solutions are sought only in the efficient set and that guidance is provided to choose the bounds to enhance solvability. To minimize $\mathbf{f}(\mathbf{x}) = (f_1(\mathbf{x}), f_2(\mathbf{x}), \ldots, f_q(\mathbf{x}))^T$, the basic trade-off algorithm is :

Choose a base point reference level as $\mathbf{a} \in \Re^q$ and a search direction as $\mathbf{b} \in \Re^q$ where $b_i \leq 0$ $i = 1, 2, \ldots, q$. Introduce a scalar parameter p and solve the following problem :

$$p_1 = \min_{\mathbf{x} \in \Re^n} \{p : \mathbf{f}(\mathbf{x}) \leq \mathbf{a} + p * \mathbf{b}\} \qquad (4.3.1)$$

p is then a measure of the amount of improvement in the "worst" f_i from $\mathbf{a}$ in the direction of $\mathbf{b}$. Define threshold level vector

$$\mathbf{a}^{(0)} = \mathbf{a} + p_1 * \mathbf{b} \qquad (4.3.2)$$

The following iterations will then generate an efficient solution from the threshold level vector :

Step k $(k = 1, 2, \ldots, q)$

(k1) Solve the problem

$$\min_{x \in \mathbb{R}^n} \left(f_k(x) : f_i(x) \leq a_i^{(k-1)} \quad \text{for} \quad i = 1, 2, \ldots, q \quad \text{and} \quad i \neq j \right) \qquad (4.3.3)$$

Denote the solution by $x^{(k)}$.

(k2) Set $a^{(k)} = f(x^{(k)})$.

The initial formulation (4.3.1) is essentially a weighted distance function formulation (3.7.1.3) with $w = b$ and $y^* = a$ while the subsequent formulation (4.3.3) is essentially the ϵ-constraint formulation with the successive execution of the auxiliary optimizations to ensure solution efficiency. It is also pointed out that it is only in the rare case when the efficient set in the performance index space contains straight line segments parallel to some f_i-axis, that more than one step is required.

By choosing different combinations of the a and b vectors, different efficient solutions can be obtained.

Fleming et al. [Fle 86] exploited the fact that (4.3.1) is essentially a goal programming formulation, which is equivalent to the weighted distance function (3.7.1.3a). The base point reference levels are set to desired levels, called goals. Elements of the design direction b are interpreted as weights for the indices. They provided guidance to help choosing them for an initial design :

Rule 1 : Set b_i to 0 for those indices which goals must be realized;

Rule 2 : Set the remaining b_i to the corresponding a_i's.

Rule 1 establishes hard constraints and rule 2 assigns equal importance to the achievement of others.

A distinguished approach is due to Nye and Tits [Nye 86]. Their formulation is

$$
\min_{x \in X} \quad \max_{i} \left\{ \bar{f}_i = \frac{f_i(\mathbf{x}) - f_i^g}{f_i^b - f_i^g} \quad i = 1, 2, \ldots, q \right\}
$$
$$
X = \{ \mathbf{x} : \mathbf{x} \in \Re^n, \quad g_j(\mathbf{x}) \le 0 \quad j = 1, 2, \ldots, r \}
$$

$$(4.3.4)$$

where f_i^g and f_i^b are the assigned good and bad values for index i. By assigning the good and bad values for each index, the designer effectively scales all indices to a common scalar measure of satisfaction. The formulation (4.3.4) then seeks to improve the least satisfactory index at each iteration. This is essentially a weakly monotonic order-representing achievement function (3.7.1.4c). However, it uses both the reference point $\bar{y}$ and the vector of weights $\mathbf{w}$ of the weighted distance function F_{d_2} in (3.7.1.3b) as the scalarizing parameters, giving extra freedom to the designer. They employ semi-infinite optimization algorithms to solve (4.3.4) and indeed, the original intention was to enhance the usability of semi-infinite optimization by a multi-objective formulation [Nye 83]. They also observe that the functional constraints are often soft and should be included in the performance indices f_i rather than the hard constraints g_i as in the formulation (4.2.1).

Jacob and Deng used weighted sum of the indices instead of goal programming for generating efficient solutions [Jac 86].

Gopalsami and Sanathanan take the satisficing approach [Gop 85]. They suggest selecting a particular efficient solution based on the size of its neighbourhood in the parameter space which also satisfies the desired levels for the indices. For simplicity's sake, the largest hypersphere contained in the neighbourhood set instead of the set itself is to be maximized.

Evaluation

The evaluations of different scalarizations in section 3.7.1 can be carried over to those of this section. However, the formulation (4.3.4) due to Nye and Tits seems to be providing very meaningful scalarizing parameters to the designer. We can expect him to be very comfortable in expressing his wishes in terms of good and bad values for indices. However, this is achieved by using both w and $\bar{y}$ which requires the designer to handle $2q$ scalarizing parameters. If q is large, the cognitive load required for the expression of wishes may be too much out of proportion to the importance of this activity among others.

4.4 Performance Vector Optimization

Unlike the methods cited in the previous section, performance vector optimization is a true IMOP approach with an emphasis on the designer-computer interaction. It aims to improve the performance in an incremental manner with rapid interaction. Designs are frequently evaluated and the optimization is closely guided by the designer.

This method was developed in a practical design environment (DFVLR in West Germany [DFV 84]) to solve the class of multi-model/multi-criteria problems. "Multi-model" refers to the approach of using a set of linear systems to represent inherently non-linear or uncertain linear systems for the purpose of robust control system design. An interactive FORTRAN program REMVG is the devoted software implementation of the method.

Unlike Zakian's emphasis on a careful choice of performance indices, this approach aims to deal with complex design problems using a liberal choice of performance indices. The emphasis is on aiding the designer to handle them and seeking incremental improvements in a highly interactive manner with systematics and efficiency. Design proceeds as iterations of the following substitute scalar optimization problem:

$$\min_{x \in \mathbb{R}^n} F^v(x) = \max_i \left(\bar{f}_i(x) = f_i(x)/c_i^v \quad i = 1, 2, \ldots, q \right) \tag{4.4.1}$$

where f_i's are the performance indices which are restricted to be non-negative, with low values preferred. The c_i^v's are bounds for the f_i's which the designer assigns at the v'th iteration such that

$$f_i(x^{v-1}) \leq c_i^v \leq c_i^{v-1} \qquad i = 1, 2, \ldots, q \tag{4.4.2}$$

Enforcing such a relation guarantees the following property :

$$f(x)^v \leq c^v \leq c^{v-1} \leq \ldots \leq c^1 \leq c^0 \tag{4.4.3}$$

i.e. subsequent design x^{v+1} will have its performance index vector $f(x^{v+1})$ bounded by one more vector $c^{v+1},\ldots$ etc (fig. 4.3). Therefore, the search can be visualized as a successive pruning of undesirable regions in the index space. If it is suspected that a desirable region has been pruned due to a combination of the bound vectors, the iterations can be restarted with a new c^0.

It is advised to take

$$c_i^v = c_i^{v-1} \tag{4.4.4}$$

for those indices whose values are small enough and

$$c_i^v = f_i(x^{v-1}) \tag{4.4.5}$$

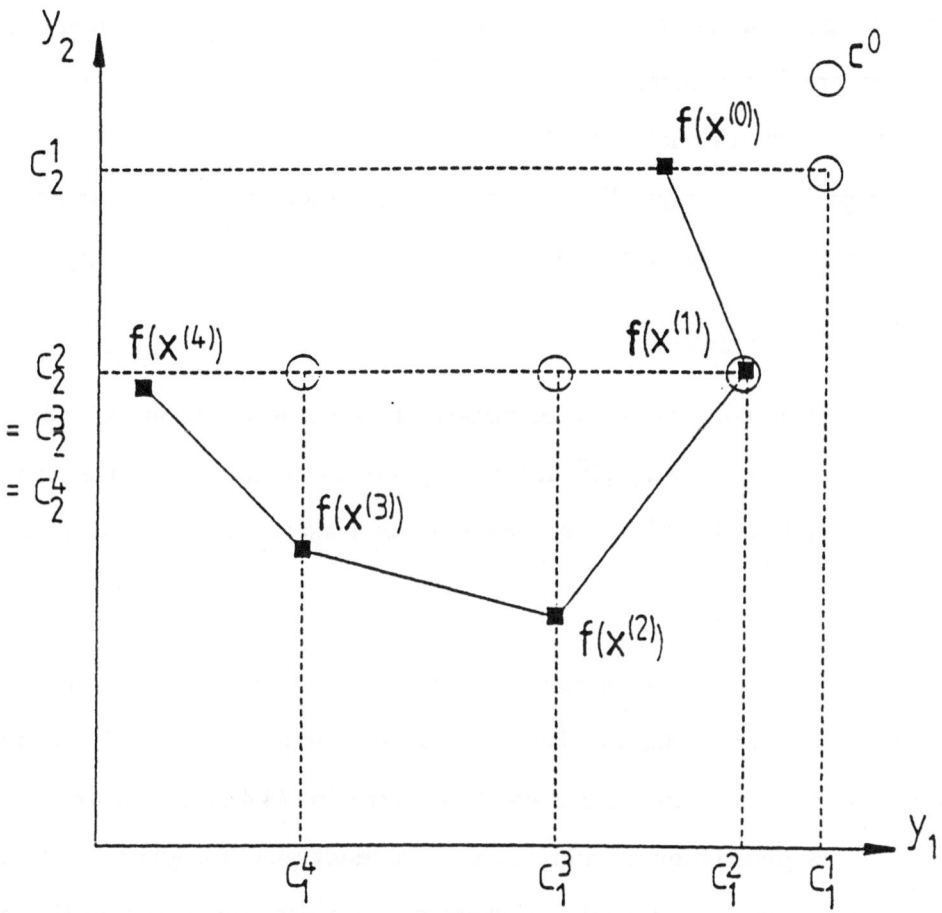

$$c^0 \geq c^1 \geq c^2 \geq c^3 \geq c^4 \geq f(x^{(4)})$$

4.3 Performance Vector Optimization

for those whose values are to be improved further. Solving such a minimax optimization problem changes the values of all indices to achieve uniform maximum percentage reductions relative to c_i^v's. It is stressed that the minimization is a tool and not a goal of the design. Formulation (4.4.1) are solved only to an approximate extent by setting an upper bound for the number of function evaluations for each iteration v.

This is essentially the weighted distance function formulation (3.7.1.3b) with $\bar{y} = 0$ and $\mathbf{w} = (c_1, c_2, \ldots, c_q)^T$ and is very similar to the interactive satisficing method, which is an IMOP method developed by Nakayama et al. [Nak 84].

Evaluation

This approach is developed primarily for implementation as a practical CACSD procedure. Careful consideration has been put on practicability, usability and natural integration into the design process. The restriction (4.4.2) is imposed for more systematic designer-computer interaction. As a result (4.4.3) is guaranteed, which can be interpreted as some kind of convergence to a desirable region in the index space.

Such an approach results in a very appealing generator. The strength lies in the expressive power and control made available to the supervisor in the assignment of the c_i^v's. He can control the design effort in each iteration v by specifying the maximum number of function evaluations to be carried out. Using a large number of indices also enable him to express the original design specifications with precision. For instance, when designing a multivariable controller, he can express his wish in reducing the interaction in a particular channel to be the only improvement that he is seeking. Fine control over the design direction in the performance index space is achieved by solving each formulation to an approximate extent in favour of more frequent adjustment of the formulations themselves to express design wishes with better precision.

As contrast to the precision of minimization problems that semi-infinite optimization offers, this precision in the expression of design wishes is the precision required. He will find it relatively easy to monitor changes in the substitute problem formulations (4.4.1) so as to enhance correspondence with the original design specifications as he proceeds.

In the REMVG implementation, an evaluator is also explicitly constructed and tailored to support the intelligent supervisor's decision making. The designer assigns a set of analysis procedures to be automatically conducted on the new candidate at the end of each iteration. In addition, the performance vector and parameters of the new candidate are displayed alongside those of the previous one for comparison to support the designer's next choice of c according to (4.4.2). In this way, the information flow across the modules is being co-ordinated for its effective use.

This approach encourages design precision with a large number of indices. Frequent interaction to evaluate candidates and modify the formulation enables a rapid prototyping of candidate designs and a good control over the design direction in the index space. The designer is encouraged to frequently evaluate and compare candidates and articulate his wishes with accuracy and flexibility. The quality of cooperation between the designer and the computer guaranteed is unsurpassed among the other approaches we have cited in this section. We therefore give much merit to this approach which demonstrates the power of an intelligent application of IMOP to CACSD.

CHAPTER 5

A STRATEGY FOR CACSD

A strategy is a plan to identify and deploy resources for resolving a problem situation. In the previous chapters, we have developed a conceptual view for the designer and a framework for the computer to ensure cognitive compatibility between the designer and the computer for a good quality of interaction. We have also surveyed the areas of IMOP and search-based control system design methods. We now propose an effective strategy for the designer to use these and other design facilities made available in the computer to resolve control system design problems, in other words, a strategy for CACSD.

We first discuss the transcription of a CACSD problem into one of IMOP in the following section. The strategy has as its objectives a set of design principles which are described in section 5.2. The strategy itself is a two-level plan for the design process reported in the last section.

5.1 CACSD as IMOP

To transcribe a CACSD problem into one of IMOP, a set of performance indices f_i, $i = 1, 2, \ldots, q$ are selected to pose it as the vector minimization problem (3.2.4). A feasible set X has to be given as the set of controllers to be searched. Without loss of generality, we can have $X \subset \Re^n$, a subspace of the n-dimensional Euclidean space (we can always shift among subspaces of different dimensions as often as we like if such is necessary to represent different design methods represented by the generalized search). We note the following characteristics of the resulting IMOP problem :

(i) the f_i's are non-linear maps in general. This characteristic should discourage the occurrence of the exceptions (3.3.5).

(ii) there are intrinsic trade-offs among the common control-theoretic objectives, e.g. the well-known trade-off between performance and control cost.

(iii) It is unusual to generate a design dominated by a previously obtained one with respect to all performance indices, *and vice versa*.

The last point can be appreciated as follows. The dynamical nature of control systems and the plants often necessitates the use of trajectories for evaluations, e.g. time-domain trajectories such as step responses and frequency-domain trajectories such as open-loop bode plots. The trajectories are essentially infinite-dimensional. Also, intrinsic trade-offs often exist within one trajectory. A step response with a fast rise time tends to overshoot. The open-loop bode plot carries the trade-off between gain manipulation for performance and phase requirement for stability. Therefore, the performance indices which evaluate these trajectories tend to carry such trade-off relationships always, which contributes to some of the intrinsic trade-offs.

This last point has an important implication. The effective dimension of Y, the image of the feasible set, is often smaller than q. In such a case, the dimension of the efficient frontier Y^* in general is the same as that of Y and the efficient set X^* often assumes full dimension n. Therefore once X^* is hit (an efficient solution is found), only moderate amount of effort is required to keep the search within it.

From these observations, we derive two working assumptions :

[WA1] All Efficiency

There is no distinction between weak efficiency and efficiency. All scalarizations in section 3.7.1 are therefore sufficient for efficiency in this situation.

[WA2] Efficient Trajectory

If an efficient design is used as the initial point in an auxiliary optimization of a scalarization which is sufficient for efficiency, most, if not all, of the subsequent trajectory of iterated points are efficient.

5.2 Design Principles

We have derived a set of design principles for our design approach. They provide the overall objectives of the strategy.

[DP1] Pluralistic Use of Design Methods

It is generally accepted that no single design method is good for all situations. The choice depends on the problem's structure and the resources available. However, when design methods are seen as generators *in co-operation with the other modules*, this choice should become more pluralistic. Different methods should then be employed to generate more candidates.

[DP2] Uniform Trade-offs and Evaluations

While the designer may use more than one design method, the resulting designs should be evaluated and trade-offs executed with respect to the same design specifications in a uniform manner. Although the specifications may be modified, they should not create any bias among candidate designs generated from different methods.

[DP3] Rapid Prototyping / Empirical Learning

Candidate designs are to be generated at a high rate so that whenever a new candidate is generated, many previous candidates are still in sight. This should encourage a holistic view for evaluation and trade-offs. The short term memory of the designer can be better utilized. However, some continuity of the performance of the series of candidates is important. Also, with a large number of candidate designs, analysis on the resulting sample *at large* may reveal useful information to the designer.

[DP4] Design Granularity

A liberal choice of performance indices should be used to represent the design specifications. Precision of representation may then be achieved, resulting in a high resolution in the comparison and evaluation of candidate designs as well as a more accurate expression of designer's wishes.

[DP5] A Priori Search Control

The designer should seek to control the outcome of a generator's automated search *a priori*. This can be achieved when the generator searches by an explicit IMOP using a scalarization with meaningful parameters (meaningful as in the sense defined in section 3.7.1). For instance, using an achievement function formulation with an attainable reference point, the solution to the auxiliary optimization problem is always an efficient solution which dominates the reference point.

[DP6] Open Design

An open design is one in whose neighbourhood, alternative designs with similar attributes are readily accessible. Such designs make possible further selection, and therefore a wider choice for the designer. This is valuable especially when there are *a posteriori* changes in the design specifications, probably after the customer has seen a suggested design. This may be achieved if the design is obtained using an iterative numerical search.

5.3 The Strategy : A Two-Level Plan of the Design Process (Fig. 5.1)

As we have noted before, most CACSD methods contribute as generators in our conceptual view of GCS (and therefore, to the tackling of the design generation sub-problem). Requirements of the other sub-problems of CACSD as represented by the other modules of GCS (viz., design data management in the information handlers, design evaluation in the evaluators, decision making in the intelligent supervisor) often receive only minimal attention. This may be justified since these requirements are not particular to control system design and therefore should be supported instead by the CAD community at large. Much of the existing computing support for CACSD manifests such a philosophy [Lit 84], [Mat 87].

However, the success of vector performance optimization has demonstrated the value of an integral support of these requirements with an intelligent application of IMOP. A versatile design approach results, which incorporates some of the designer principles, viz. [DP2], [DP3], [DP4] and [DP5]. However, the method itself is not a complete design approach. Broadly speaking, there are two aspects of practical design: innovation and trade-offs. This approach is essentially executing iterations of trade-offs while assuming innovated initial designs. In the following, we propose a strategy for CACSD as a two-level plan, which co-ordinates both the innovation and the trade-off aspects. It incorporates all design principles cited in the previous section and pays due attention to the requirements of all the modules in GCS. The distinctive features of the two levels are (i) the innovation of a parametrized feasible set in level I and (ii) the maintenance of a sample efficient subset in level II.

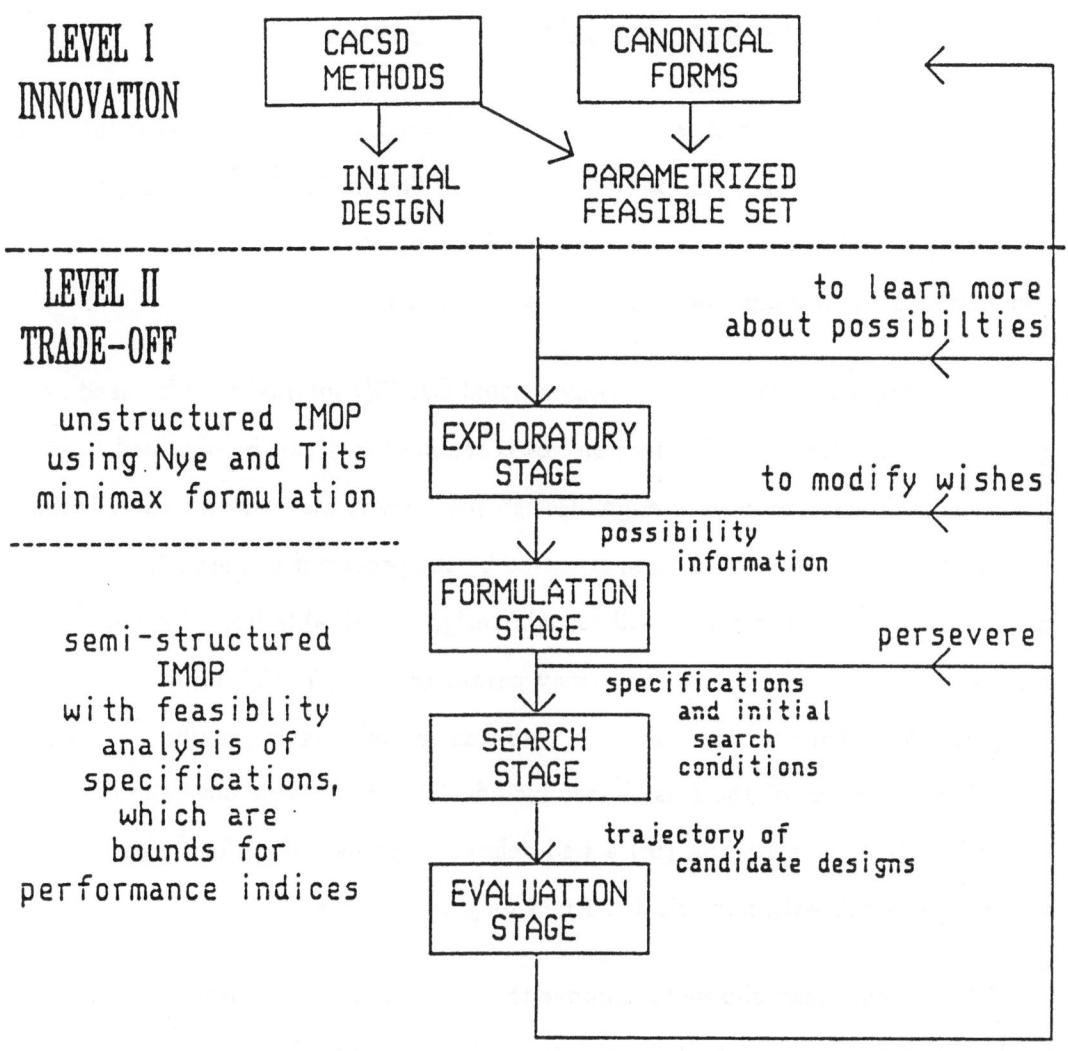

LEVEL I INNOVATION

CACSD METHODS

CANONICAL FORMS

INITIAL DESIGN

PARAMETRIZED FEASIBLE SET

LEVEL II TRADE-OFF

to learn more about possibilties

unstructured IMOP using Nye and Tits minimax formulation

EXPLORATORY STAGE

to modify wishes

possibility information

FORMULATION STAGE

persevere

semi-structured IMOP with feasiblity analysis of specifications, which are bounds for performance indices

specifications and initial search conditions

SEARCH STAGE

trajectory of candidate designs

EVALUATION STAGE

5.1 The Design Strategy : A Two-Level Plan of the Design Process

5.3.1 Level I : Innovation - Feasible Set Generation

The aim of this level is to innovate an initial design *as an element of a parametrized feasible set*, i.e. $x^{(0)} \in X \subset \Re^n$. The initial design acts as an entry point into the feasible set for the auxiliary optimizations in the next level. Innovation is the principal occupation of the designer.

To obtain an initial design, a conventional CACSD method may be used as a generator. To enter level II, a feasible set is required and can be obtained from a parametrization of the initial design. Design parameters may be constructed from the substitute problem formulation accepted by the generator of the design method used. The resulting feasible set will contain designs obtainable from this particular method. Alternatively, the designer may parametrize the initial design according to a canonical form. In this case, the resulting feasible set is usually very large. Desirable properties of the feasible set include (i) a reasonable coverage of good controllers, (ii) convexity and (iii) a well-behaved topology of performance so that subsequent search with numerical routines may be effective.

The designer may choose to improve the initial design while remaining at this level. However, as soon as the resolution of conflicting objectives becomes the dominant activity, level II should be entered to handle the trade-offs required to match his wishes with the design possibilities. Therefore, level I may be executed with any design method (design principle [DP1]) which gives a reasonable feasible set while level II provides a uniform way to exercise trade-offs ([DP2]).

5.3.2 Level II : Trade-offs - Matching Wishes and Possibilities [Ng 87]

The aim of this level is to match the designer's wishes with design possibilities of the feasible set with the help of IMOP. Performance indices are identified. The efficient frontier in the resulting index space is then sought and auxiliary optimizations of scalarizations are used to search *and sample* the frontier. The search continues until a satisfactory region is located, whilst a sample efficient subset is maintained to store up all efficient candidates encountered *in the trajectories of the iterative searches.* If such a region is not found, the designer may suspect limitations from the feasible set X in that a satisfactory region may exist outside the range of X. An alternative parametrization may then be sought to expand the feasible set or level I may be re-entered to innovate an alternative feasible set. Design principles [DP3] and [DP6] may be achieved with the sample efficient subset being continually updated and interrogated.

This level is divided into four stages described as follows.

(1) Exploratory Stage

This stage aims to support the design principle [DP3] by exploring the efficient frontier in the index space with unstructured IMOP. Auxiliary optimizations of IMOP are conducted on the IMOP scalarizations. Scalarizing parameters are used to control the direction of the trajectory traversed by the iterative optimization. The designer is expected to interact frequently with the optimization by altering the scalarizing parameters and examining the trajectory.

For the purpose of exploration, a small set of aggregate performance indices $\hat{f}_i$, $\hat{i} = 1, 2, \ldots, \hat{q}$ should suffice to approximate the design specifications. This will ensure that the designer will not be overloaded with too many scalarizing parameters to handle. We suggest using Nye and Tits' minimax formulation (4.3.4). The designer should aim to explore the best possible achievement of each index, with the others within reasonable values by assigning appropriate good and bad values for the indices. We suggest $\hat{q} \leq 5$ so that the designer is handling at most ten parameters.

To ensure that the trajectory is within the efficient frontier, it suffices to solve the optimization of one scalarization exactly. The resulting design will be efficient (according to working assumption [WA1]) and the subsequent trajectory will stay on the frontier ([WA2]) for most, if not all of the time. Therefore, as the designer explores the frontier, the computer registers *all* the candidate designs encountered on the trajectory.

Such handling of the indices, each by two scalarizing parameters, should have given the designer a good understanding of the conflicts among them. A more elaborate set of indices $f_i \quad i = 1, 2, \ldots, q$ are then chosen to represent the design specifications with precision to achieve the design principle [DP4]. The candidates obtained in this stage are then evaluated for the new set of indices. A sample efficient subset is then extracted by discarding those which are disproved of efficiency (when they are dominated by some other members of the subset). This subset will be maintained and updated throughout the subsequent search which continues to register regions in the index space as well as to help the designer in narrowing down to a final design.

We note in passing that an even distribution of designs on the trajectory is desirable for the sample to be representative of the regions traversed. This poses special requirements on the numerical optimization method and will be discussed in chapter 6.

(2) Formulation Stage

The designer comes to this stage for a precise formulation of the design problem which is attainable. Based on the sample subset, goals are specified in terms of bounds for the elaborate set of indices. During these specifications, the sample subset is called upon to test the attainability of the goals by revealing any critical conflicts among them. Goals should then be specified with high attainability to achieve the principle [DP5].

Therefore, we require tools to (1) analyse the feasibility of a set of index bounds and (2) help trade-off the bounds to match the designer's wishes with design possibilities. These tools are discussed in chapter 8.

(3) Search Stage

Having assigned bounds on the indices, we come to this stage to solve the resulting set of inequalities. A design is chosen from the sample subset as the initial point. Either the moving boundary process of the Method of Inequalities or an auxiliary optimization of an achievement function formulation may be used. The moving boundary process is desirable for its ability to preserve satisfied bounds while the achievement function formulation has the advantages as stated in section 3.7.1.

The search stage either succeeds or fails in finding a solution. However, having appreciated the attainability of his goals in the previous stage, the designer may choose to re-start a failed search with a new initial point or even a different search algorithm if he has faith in the attainability of his goals.

Also, as a result of the search, more designs should have been collected and the sample subset updated.

(4) Evaluation Stage

The new trajectory of designs traversed in the search stage is evaluated against all the indices. Compared with the designs in the original sample efficient subset, new possibilities or conflicts are revealed. As a result, the designer may go back to the formulation stage for another set of goals, select a new initial point and subsequently repeat the search stage. Or another exploratory stage may be invoked for better understanding. Or an acceptable design may have been obtained and the design process terminated.

5.3.3 Discussion

Level II is often more important in practical design when the specifications are required to be manipulated *to compromise requirements with technology.* This manipulation is often the central activity of design, manifested as trade-offs in the design process. Level II provides the interface between the technology (X and $f(x)$) and the requirements (whereabout in Y is desirable) and forms the basis of the dialogue between the designer and the customer. The customer is then encouraged to play a more important role in the manipulation than is generally the case as the technology is often too remote from him.

Employing IMOP as a formal framework for CACSD, level I is incorporated in the Initialize step of IMOP's basic algorithm while level II is supported as iterations of the AN-Generate and DM-Trade-off steps. While conventional computing support for CACSD often pays undue attention to level I, the IMOP framework and the two-level plan put the proper emphasis on the trade-off aspects of practical design.

CHAPTER 6

NUMERICAL SEARCH TOOLS

We describe a set of tools for the generation of candidate solutions in the IMOP of the design strategy. There are two basic IMOP problems in the strategy. In the exploratory stage, Nye and Tits' formulation (4.3.4) is used and in the search stage, it is either Zakian's inequality formulation (4.1.1) or Wierzbicki's achievement function formulation (3.7.1.4). In all cases, iterative numerical search methods are used.

We shall denote the candidate generated in the k'th iteration by $x^{(k)}$. The initial point is $x^{(0)}$ and iterations start with $k = 1$.

6.1 Candidate Solution Generation in The Exploratory Stage

The formulation is one of constrained minimax optimization :

$$
\min_{x \in X} \quad F(x) = \arg\max \left\{ \bar{f}_i(x) = \frac{\hat{f}_i(x) - \hat{f}_i^g}{\hat{f}_i^b - \hat{f}_i^g} \quad \hat{i} \in \hat{Q} = \{1, 2, \dots, \hat{q}\} \right\}
$$

$$
X = \{x : x \in \Re^n, g_j(x) \le 0, \quad j \in R = \{1, 2, \dots, r\}\}
$$

(6.1.1)

where $\hat{f}_i^g$ and $\hat{f}_i^b$ are the assigned good and bad values for the aggregate performance index $\hat{f}_i$.

We note the following features and requirements in solving this problem in the exploratory stage :

(i) It is a non-smooth optimization problem. In general, $F(x)$ is non-smooth whenever $\bar{f}_i(x^{(k)}) = \bar{f}_j(x^{(k)})$ for some $\hat{i}, \hat{j} \in Q, \quad \hat{i} \ne \hat{j}$.

(ii) The formulation is *ad-hoc* in the sense that the aggregate performance indices are constructed in a highly problem-dependent way with no general forms and analytic expressions for the derivatives are not available in general. Therefore, we shall seek a reasonable search algorithm which is as *general-purpose* as possible.

(iii) The formulation is not so much an optimization problem to be solved exactly than a vehicle for sampling the efficient set. Therefore, even if available, "good" optimization algorithms with rapid convergence properties may not be desirable. Also, such algorithms often require the approximation of derivatives by finite differences. When the design parameter space is of very high dimension, most of the time will be spent in such approximation. It may be more desirable to divert such effort to sampling.

(iv) The search is to be conducted in a highly interactive manner. The collection of data which fully defines the state of an iterative search should be simple to enable the designer's analysis and interrogation.

(v) The progress of the search is important to the designer and should be closely monitored for the variation of the individual aggregate performance indices. These will help identify the convergence of a search and the cause of it.

6.1.1 Simplex Polytope Direct Search Method

In regard to the above considerations, we have chosen the direct search method of Nelder and Mead [Nel 65] which conducts a local search by maintaining a simplex polytope with $(n + 1)$ apexes in the n-dimensional design parameter space. The simplex polytope undergoes geometrical transformations such as reflections, expansions and contractions according to a set of transformation rules as the means to locate the minimum (fig. 6.1). This direct search method is very general-purpose and is well-known for its robustness in its progressive approach to the minimum. Other direct search methods such as Powell's direction set method [Pow 64] and Rosenbrock's derivative-free method [Ros 60] locate the minimum by seeking descent directions and conducting linear searches along them. They often fail at the non-smooth points when no descent directions are identified among a discrete set of directions they maintain (fig. 6.2) while the simplex polytope will continue its various transformations to seek the descent directions.

The orientation of the polytope fully defines the state of the iterative search and its examination will be meaningful to the designer as a cluster of neighbouring points. It therefore would be suitable to be used in a highly interactive manner. If the apexes of the polytope at the k'th iteration are $z_m^{(k)}$, $m = 1, 2, \ldots, n+1$, we denote the polytope by

$$Z^{(k)} = \left(z_1^{(k)} \quad z_2^{(k)} \quad \ldots \quad z_{n+1}^{(k)} \right) \tag{6.1.1.1}$$

The $(k+1)$'th iteration simply maps $Z^{(k)}$ to $Z^{(k+1)}$. The designer can affect the search by replacing a point in $Z^{(k-1)}$ with one of his choice or even initiating a new polytope.

The polytope method is intended for unconstrained optimization. There are extensions to handle the constrained case, e.g. the Box method [Box 65] and the penalty method [Maz 87]. They maintain polytopes with more apexes and more complicated transformation rules to guard against the dimension collapsing problem, especially at constraint boundaries (fig. 6.3), which is the major reason for any failure of the polytope method to locate a local minimum.

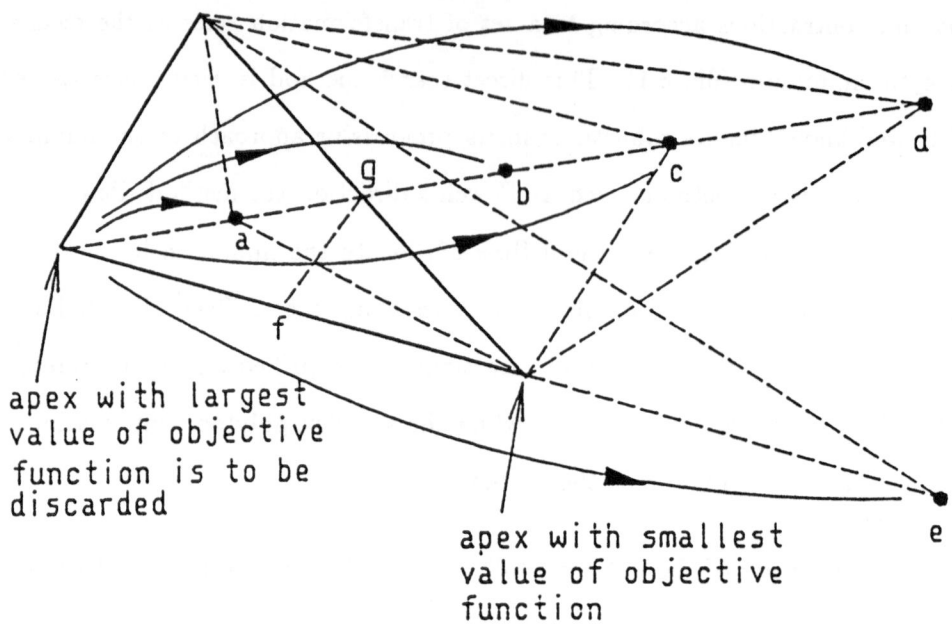

apex with largest
value of objective
function is to be
discarded

apex with smallest
value of objective
function

TRANSFORMATIONS

a   contraction
b   contraction
c   reflection
d   expansion
e   reflection
f, g comprehensive contraction

6.1 Geometrical Transformations of the Simplex Polytope

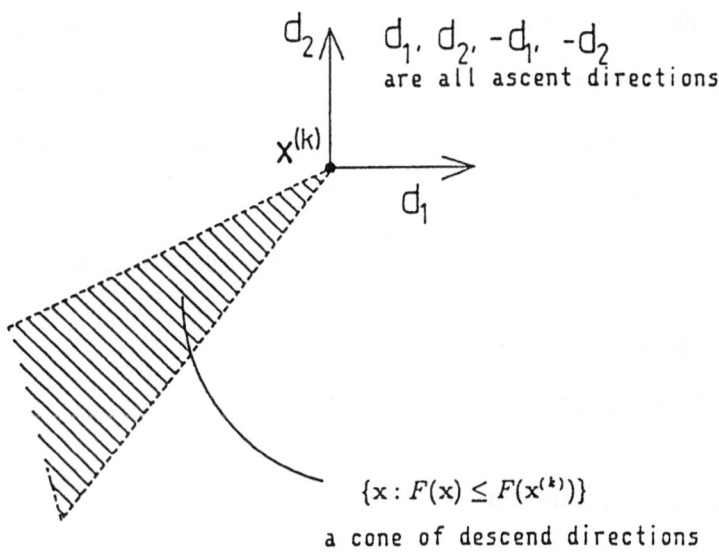

$\{x : F(x) \leq F(x^{(k)})\}$

a cone of descend directions

6.2 Failure of Some Direct Search Methods at a Non-smooth Point

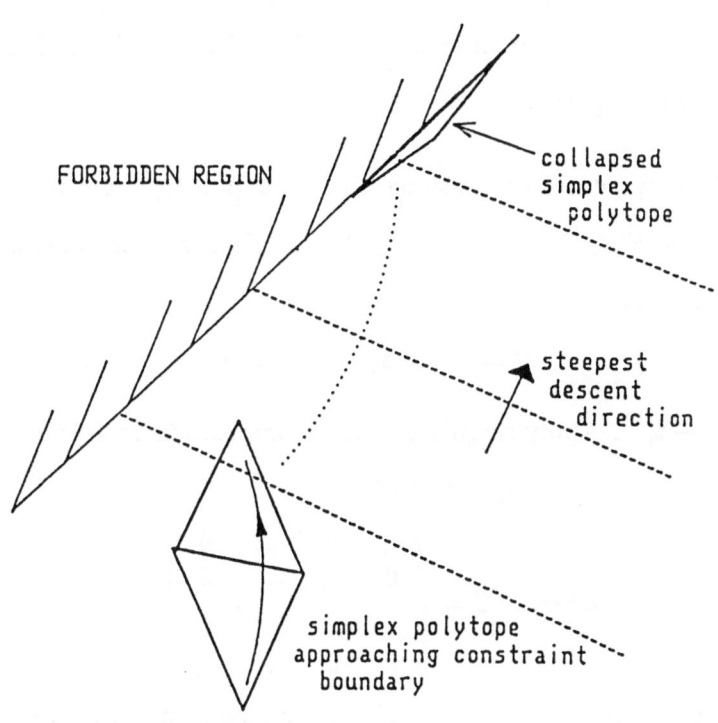

6.3 Dimension Collapsing

We have chosen a simpler alternative without increasing the number of apexes by employing special procedures to restore full dimensions for collapsed polytopes. We also retain an unconstrained problem by including the hard constraints g_i into the domain of the *max* operation, i.e.

$$\min_{x \in \mathbb{R}^n} \quad F(x) = \arg\max_{i,j} \left\{ \{\bar{f}_i(x), \quad \hat{i} \in \hat{Q}\} \cup \{\bar{g}_j(x), \quad \hat{j} \in R\} \right\}$$

$$\text{where} \quad \bar{g}_j(x) = \frac{g_j(x) + \delta}{\delta} \qquad (6.1.1.2)$$

and the positive parameter δ is set to a sufficiently small value.

6.1.2 Dimension Collapsing

A trivial way to restore a collapsed polytope is simply to create a new polytope with full dimensions. However, this would destroy any valuable information about the local topology captured in the current polytope. To guard against dimension collapsing while preserving such information, we monitor the dimensionality of the polytope in the following way.

Without loss of generality, assume that the polytope apexes are ordered such that $F(z_1^{(k)}) \leq F(z_2^{(k)}) \leq \ldots \leq F(z_{n+1}^{(k)})$ (therefore, $z_1^{(k)}$ achieves the minimum among the apexes). Let

$$D = (\, d_1 \quad d_2 \quad \ldots \quad d_n \,) \qquad (6.1.2.1)$$

where $d_l = z_{l+1}^{(k)} - z_l^{(k)}$. Apply singular value decomposition to D, i.e.

$$D = V \Sigma U^T$$

$$\text{where} \quad \Sigma = \begin{pmatrix} \sigma_1 & 0 & \ldots & 0 \\ 0 & \sigma_2 & \ldots & 0 \\ \vdots & \vdots & \ddots & \vdots \\ 0 & 0 & \ldots & \sigma_n \end{pmatrix} \qquad (6.1.2.2)$$

The $n \times n$ square matrices V and U are orthonormal matrices (therefore $V^T V = V V^T = U^T U = U U^T = I_n$) and $\sigma_1 \geq \sigma_2 \geq \ldots \geq \sigma_n \geq 0$, as is the usual convention of singular value decomposition. If $V = (\, \mathbf{v}_1 \quad \mathbf{v}_2 \quad \ldots \quad \mathbf{v}_n \,)$ and $U = \{u_{ij}\}$, (6.1.2.2) implies

$$\mathbf{d}_l = \sum_{i=1}^{n} u_{il}(\sigma_i \mathbf{v}_i) \tag{6.1.2.3}$$

The u_{il}'s may be interpreted as the components of $\mathbf{d}_l$ in the space with $\sigma_i \mathbf{v}_i$'s as a set of orthogonal basis vectors. Also, σ_i indicates the extent to which the $\mathbf{d}_l$'s span the subspace $\mathbf{v}_i$. The number of strictly positive σ_l gives the absolute dimensionality of the polytope while a very small singular value implies serious collapse of the polytope in the corresponding singular direction, viz. if σ_l, $l = n - n_s + 1, n - n_s + 2, \ldots, n$ are considered very small, the polytope is deficient in the subspace spanned by the last n_s columns of V, i.e. $\mathbf{v}_l$, $l = n - n_s + 1, n - n_s + 2, \ldots, n$.

A simple but effective "healing" procedure is to replace the very small singular values with more reasonable ones; σ_{n-n_s} should suffice. Let

$$D^N = V\Sigma^N U^T$$

$$\text{where} \quad \Sigma^N = \begin{pmatrix} \sigma_1^N & 0 & \cdots & 0 \\ 0 & \sigma_2^N & \cdots & 0 \\ \vdots & \vdots & \ddots & \vdots \\ 0 & 0 & \cdots & \sigma_n^N \end{pmatrix} \tag{6.1.2.4}$$

$$\sigma_l^N = \sigma_l \qquad l = 1, 2, \ldots, n - n_s$$

$$\sigma_l^N = \sigma_{n-n_s} \qquad l = n - n_s + 1, n - n_s + 2, \ldots, n$$

The healed simplex polytope will have as a new set of apexes

$$\mathbf{z}_m^{(k+1)} = \begin{cases} \mathbf{z}^{(1)} & m = 1 \\ \mathbf{z}^{(1)} + \mathbf{d}_{(m-1)}^N & m = 2, 3, \ldots, n+1 \end{cases} \tag{6.1.2.5}$$

where $\mathbf{d}_{m-1}^N$ is the $(m-1)$'th column of D^N.

An alternative healing procedure is simply to replace a selection of n_s columns of D by the last n_s columns of V to generate D^N. The obvious choices are the last

n_s columns of D which come from the apexes with larger values of F. However, it may so happen that replacing this set renders the resulting polytope deficient in the subspace spanned by the first $(n - n_s)$ columns of V. Their components in this subspace should therefore be checked. However, unlike the previous procedure when the singular values are determined in (6.1.2.4), those of the D^N constructed in this one are not determinate from the original ones. The advantage of this procedure is that only n_s apexes are replaced in the new polytope, compared with $(n-1)$ in the previous method. This may be important if the dimension of the design parameter space is high and function evaluations are costly.

6.1.3 Graphical Display Monitors

For convenience of notations, let $\mathcal{P}^{(k)}(\hat{Q}) = ([1], [2], \ldots, [\hat{q}])$ be a permutation of the aggregate indices in descending order of activeness, i.e.,

$$F(\mathbf{x}^{(k)}) = \hat{f}_{[1](k)}(\mathbf{x}^{(k)}) \geq \hat{f}_{[2]^k}(\mathbf{x}^{(k)}) \geq \ldots \geq \hat{f}_{[q](k)}(\mathbf{x}^{(k)}).$$

When the iterative optimization converges, there are four possible reasons :

(i) Dimension collapsing has restricted the search to a linear subspace in which the current point is minimum (fig. 6.4);

(ii) One of the most active performance indices (i.e. $[\hat{i}]$ such that $f_{[i]}(\mathbf{x}) = F(\mathbf{x})$) is converging to a local minimum (fig. 6.5a);

(iii) There are more than one most active performance indices and they are in conflict, i.e. the current point is efficient with respect to them (fig. 6.5b).

(iv) The most active indices are in conflict with some hard constraints (fig. 6.5c).

The dimension collapsing case can be monitored as in the last section. We

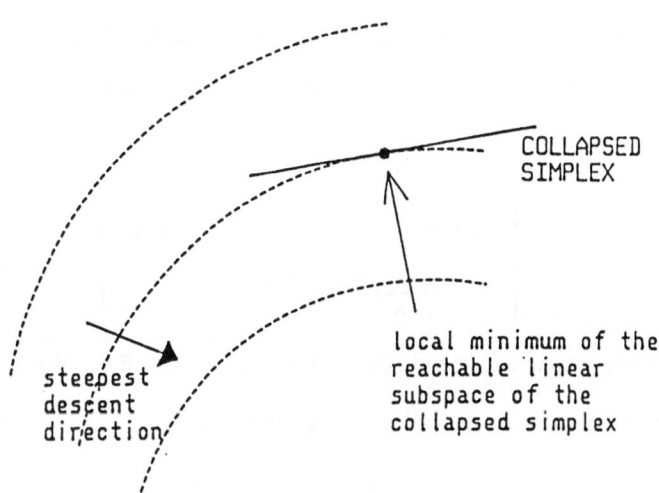

6.4 Local Minimum in a Linear Subspace

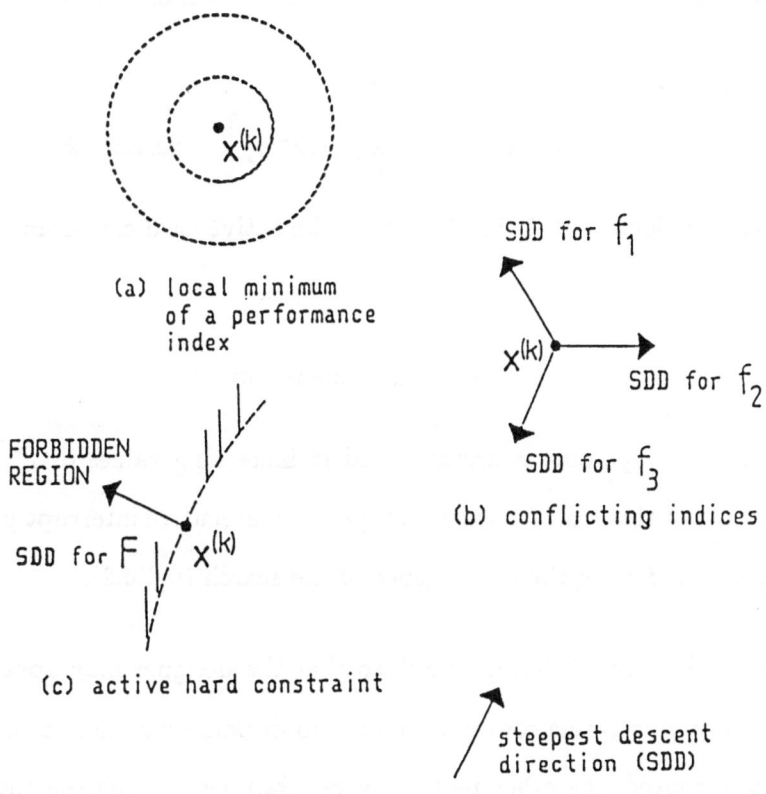

6.5 Convergence Possibilities of a Minimax Search

propose to use the following graphical displays as monitors for the iterative search with the aim of distinguishing between the other possibilities :

[P1]

$$\left\{ \begin{array}{c} \bar{f}_{[i]^k}(x^{(k)}), \quad \hat{i} = 1, 2, \ldots, n_{[P1]} \\[2mm] \max_{j \in R}\{\bar{g}_j(x^{(k)})\} \end{array} \right\} \quad \text{versus} \quad k$$

The three cases can be distinguished as in fig. 6.6. We suggest $2 \leq n_{[P1]} \leq 4$ for an informative, as well as clear, display.

[P2]

$$\left\{ [\hat{i}]^k, \quad \hat{i} = 1, 2, \ldots, n_{[P1]} \right\} \quad \text{versus} \quad k$$

This plot supplements [P1] in identifying the active indices (fig. 6.7).

[P3]

$$\left\{ j^* : \bar{g}_{j^*}(x^{(k)}) = \max_{j \in R}\{\bar{g}_j(x^{(k)})\} \right\} \quad \text{versus} \quad k$$

This plot supplements [P1] in identifying the active hard constraints (fig. 6.8).

[P4]

$$F(z_m^{(k)}) \quad \text{versus} \quad m$$

where the polytope apexes are arranged in increasing values of F. Superimposing plots of this kind for the polytopes in consecutive interrupt points may be used in monitoring the convergence of the search (fig. 6.9).

[P1] should be plotted during search so that the designer may appreciate the progress and any conceivable convergence to help deciding whether to interrupt or not. Once interrupted, the other plots may be examined to analyse the progress made. When the designer decides to change the good and bad values of the indices,

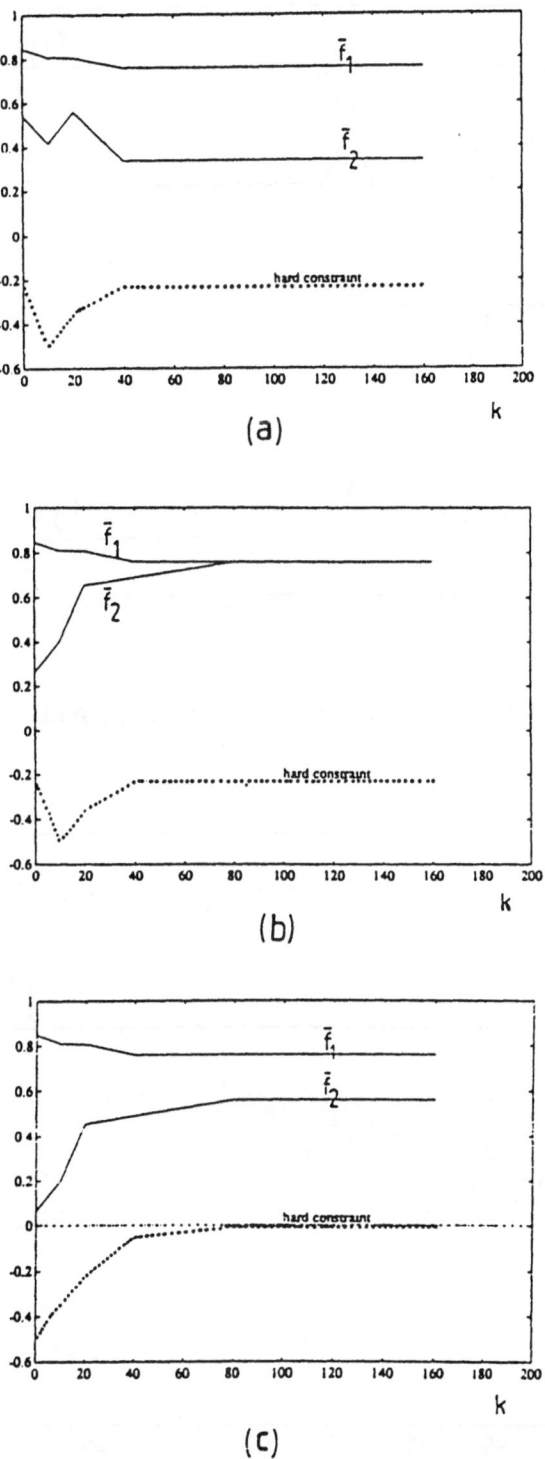

(a)

(b)

(c)

6.6 [P1]-Type Display : Distinguishing Convergence Possibilities

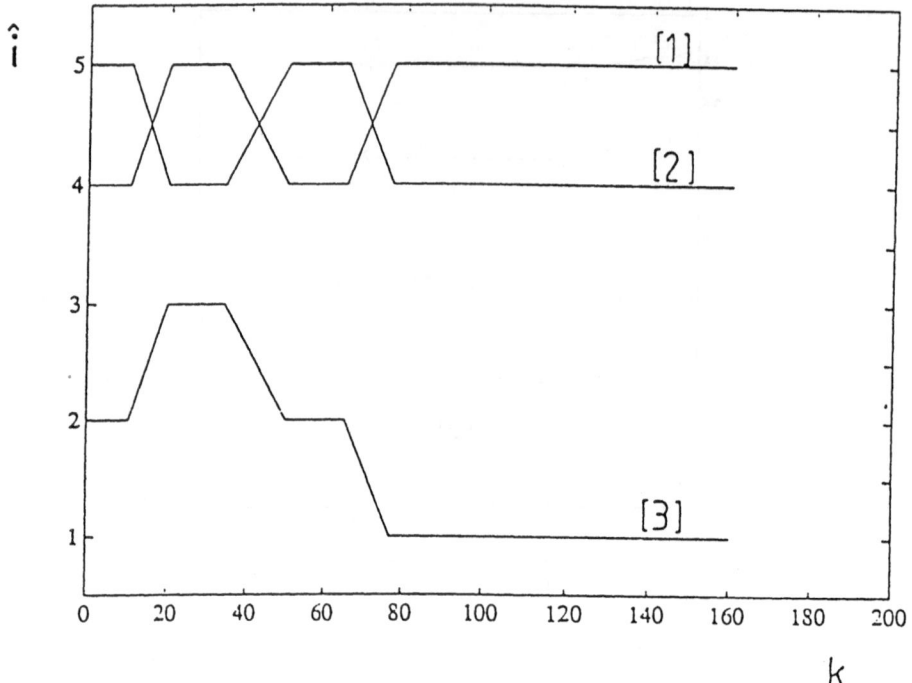

6.7 [P2]-Type Display : Identifying Active Indices

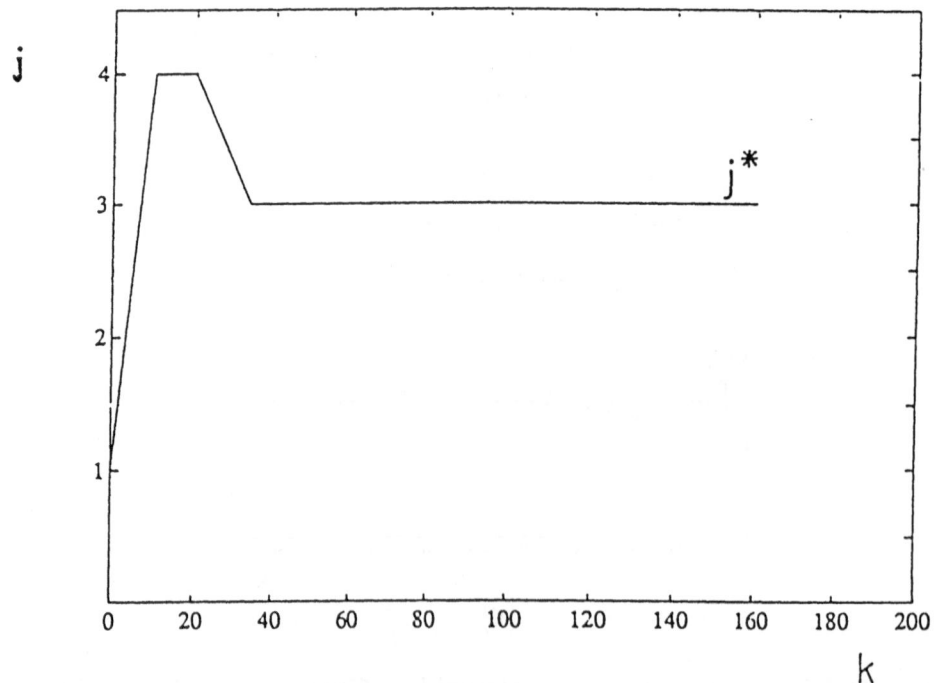

6.8 [P3]-Type Display : Identifying Active Hard Constraints

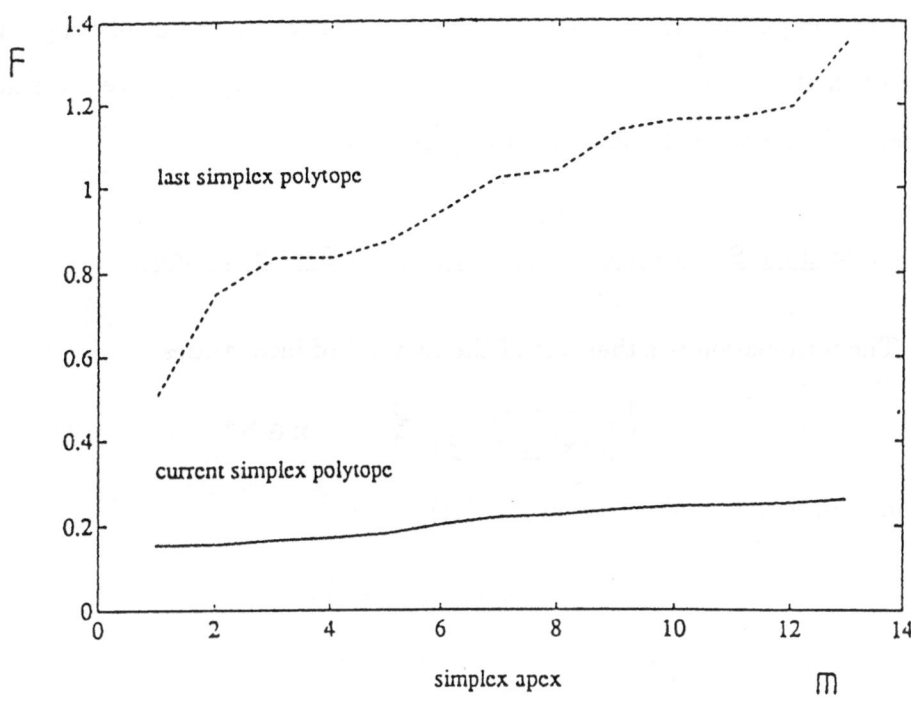

6.9 [P4]-Type Display : Monitoring Convergence in a Minimax Search

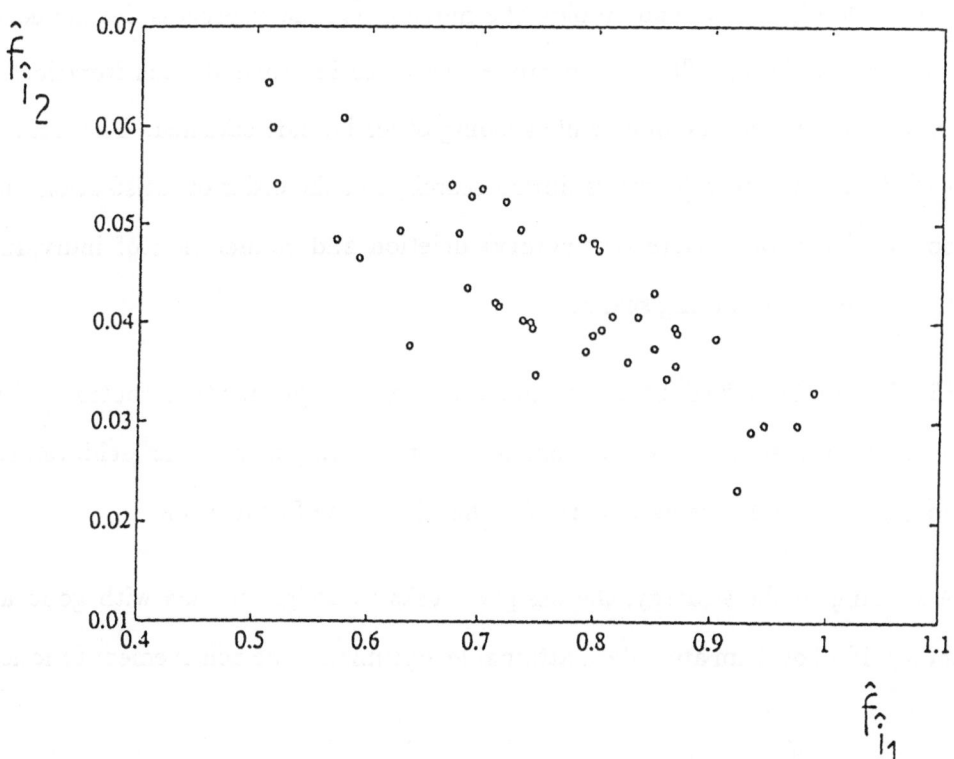

6.10 Pairwise Comparison Graph

pairwise comparison graphs of various indices, especially the conflicting ones dis-
covered in the graphs (fig. 6.10), may help in choosing appropriate good and bad
values in formulating the subsequent exploration.

6.2 Candidate Solution Generation in The Search Stage

The formulation is either one of the method of inequalities

$$\begin{cases} f_i(\mathbf{x}) \leq c_i & i \in Q \\ g_j(\mathbf{x}) \leq 0 & j \in R \end{cases} \qquad \mathbf{x} \in \Re^n \tag{6.2.1}$$

or the achievement function as in (3.7.1.4).

$$\min_{\mathbf{x} \in X} F_a\left(\mathbf{f}(\mathbf{x}) - \bar{\mathbf{y}}\right) \tag{6.2.2}$$

where $\bar{\mathbf{y}} = (c_1, c_2, \ldots, c_q)^T$.

6.2.1 Solution Generation

To tackle (6.2.1), we can employ the moving boundary process developed by
Zakian et al. [Zak 73]. The search process seeks to improve at each iteration *all*
indices with unsatisfied bounds while keeping other bounds satisfied. The number
of satisfied bounds never decreases during search, and the order of satisfaction may
be controlled by appropriate (temporary) deletion and re-inclusion of individual
inequalities in (6.2.1) during search.

(6.2.2) can be solved by an ordinary numerical optimization routine. The
optimization problem may be made easier by constructing a smoother achievement
function, e.g. (3.7.1.4c) using l_2-norm for the F_{d2} in the formulation.

According to the strategy, the designer seeks to assign bounds with good attainability. If all of them are indeed attainable, optimizing the achievement function

will guarantee the efficiency of the solution. In this sense, (6.2.2) is preferable over (6.2.1). However, the designer may choose (6.2.1) for its ability to maintain satisfied bounds while controlling the order of their satisfaction. Also, efficiency of solution is not difficult to guarantee according to the working assumptions [WA1] and [WA2] of the strategy.

In the original paper [Zak 73] when the moving boundary process was suggested, Rosenbrock's search method was used in which a set of orthogonal directions was maintained. As we discussed in section 6.1.1, it could fail to locate descent directions. We may expect this effect to be adverse in the moving boundary process since at each iteration, the search proceeds only when *all* the unsatisfied indices improve, which is possible only in a cone of descent directions (fig. 6.11). As an alternative, we can employ a *dynamic* minimax formulation such that at the k'th iteration, we solve *one step* of the following minimax problem using the simplex polytope method

$$\min_{x \in X} \quad \arg\max \left\{ \bar{f}_i = \frac{f_i(x) - f_i^g}{f_i^b - f_i^g}, \quad i \in Q \right\}$$

$$f_i^b = \begin{cases} f_i(x^{(k)}) & \text{if } f_i(x^{(k)}) > c_i \\ c_i & \text{if } f_i(x^{(k)}) \le c_i \end{cases}$$

$$f_i^g = \begin{cases} c_i & \text{if } f_i(x^{(k)}) > c_i \\ c_i^b - \delta & \text{if } f_i(x^{(k)}) \le c_i \end{cases} \tag{6.2.1.1}$$

where the positive parameter δ is set to a small value. This formulation essentially makes all indices with unsatisfied bounds equally active at the start of each iteration.

6.2.2 Graphical Display Monitors

We shall denote the set of performance indices whose bounds are satisfied by the initial point $x^{(o)}$ as Q_s. That of the unsatisfied ones is therefore $Q_u = Q - Q_s$. We propose the following graphical display for monitoring the progress of the indices in Q_s, Q_u as well as the hard constraints (fig. 6.12) :

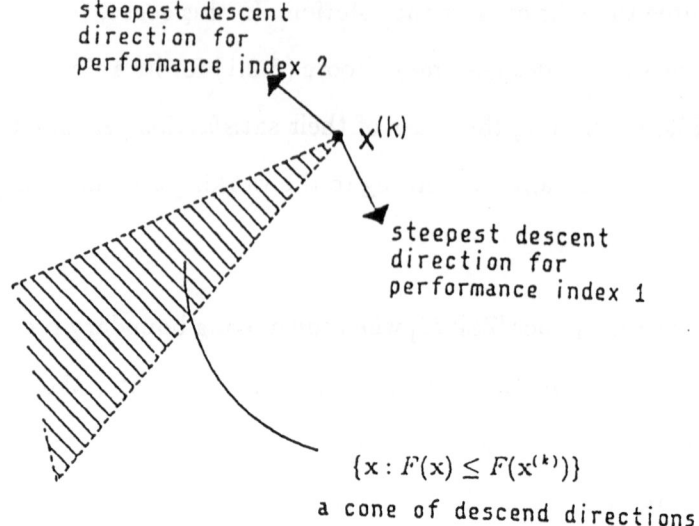

steepest descent
direction for
performance index 2

$x^{(k)}$

steepest descent
direction for
performance index 1

$\{x : F(x) \leq F(x^{(k)})\}$

a cone of descend directions

6.11 Cone of Descent Directions for the Method of Inequalities

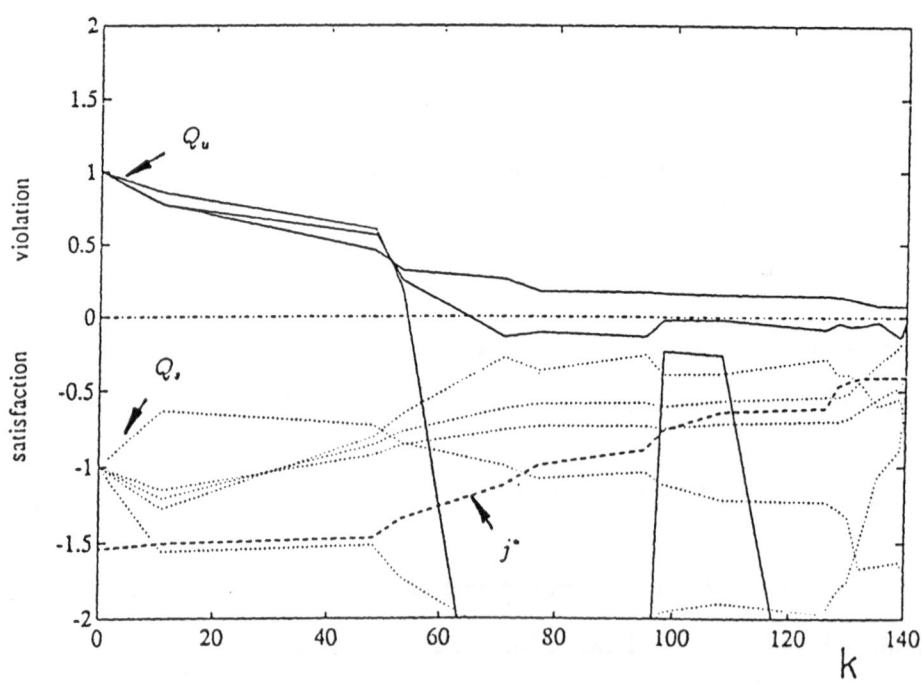

6.12 [P5]-Type Display : Monitoring Search Stage Progress

[P5]

$$\left\{\begin{array}{ll} \dfrac{f_i(x^{(k)}) - c_i}{f_i(x^{(0)}) - c_i}, & \forall i \in Q_u \\[2ex] \dfrac{f_i(x^{(k)}) - c_i}{c_i - f_i(x^{(0)})}, & \forall i \in Q_{s}, \\[2ex] \max_{j \in R}\{\bar{g}_j(x^{(k)})\} & \end{array}\right\} \qquad \text{versus} \quad k$$

The plots of all indices in Q_u start off at $(0, 1)$ while those in Q_s start off at $(0, -1)$. When the moving boundary process is used, indices in Q_u decrease monotonically while those in Q_s are always below the x-axis.

Plots similar to [P2] and [P3] in section 6.1.3 may also be used in identifying the active indices in Q_s as well as the active constraints.

CHAPTER 7

FORMULATION TOOLS

The tools to be described in this chapter help the designer to comprehend the design possibilities revealed in the sample efficient subset as well as prescribe index bounds with a view to match his wishes with the design possibilities. The general approach is one of data visualization [Tuf 83], with two-dimensional graphical displays which help the designer to visualize critical regions of the efficient frontier. These regions are those in which his wishes, expressed as bounds for the indices, are least likely to be achieved due to stringent conflicts among the indices.

7.1 Sample Efficient Subset

Denote the sample efficient subset as $X^S = \{ x_h \in \Re^n, h \in P^S = \{1, 2, \ldots, p\} \}$. Let $y_h = (y_{h1}, y_{h2}, \ldots, y_{hq})^T = f(x_h)$. Then $Y^S = \{ y_h \in \Re^q, h \in P^S \}$ is the sample efficient frontier. Assemble a data matrix for Y^S as

$$\Phi = (y_1 \quad y_2 \quad \cdots \quad y_p) \in \Re^{q \times p} \tag{7.1.1}$$

Define also $\phi_i = (\phi_{i1}, \phi_{i2}, \ldots, \phi_{ip})^T$ by

$$\Phi = (\phi_1 \quad \phi_2 \quad \cdots \quad \phi_q)^T \tag{7.1.2}.$$

As both the performance indices and the hard constraints are to be treated in the same manner, we use $Q = \{1, 2, \ldots, q\}$ to include all of them for simplicity of notation in this chapter. Therefore, the ϕ_i's are column vectors of performance index values (or hard constraint values) achieved by the p efficient candidates.

7.1.1 An Example

We shall use a sample efficient subset from an example design problem to illustrate the use of the various tools in this chapter. The design was concerned with the control of the longitudinal motion of a typical fighter aircraft represented by a linearized model [Ng 87]. A linear multivariable controller (with three inputs and three outputs) was sought. The subset comprised forty-three sample efficient candidates ($p = 43$). Thirty-four performance indices were defined and no hard constraints were formulated. Therefore, $q = 34$. We describe them briefly as follows. Indices 1, and indices 6 to 14 are concerned with control effort. Indices 2 and 15 are concerned with the structural stability. Indices 3 and 16 to 22 are concerned with performance. Indices 4 and 23 to 34 are concerned with cross-coupling between channels. Index 5 is concerned with overall system stability.

7.2 An Interactive Tool for Exploring Index Structures [Ng 88]

We shall classify the relationship between any two performance indices i, j over the whole sample subset into three kinds: viz., in correspondence ($i \mathcal{R} j$), in conflict ($i \mathcal{C} j$), or unrelated ($i \mathcal{U} j$) (fig. 7.1).

A measure of correlation s_{ij} between performance indices i, j is required to define the relations. An obvious choice is the correlation coefficient. However, a better correlation measure which is known to be more robust (in the presence of outliers or non-linearity (fig. 7.2)) is Spearman's rank correlation coefficient. To obtain the sample rank correlation matrix $S = \{s_{ij}\}$, construct $\tilde{M} = \{\tilde{\phi}_{ih}\} \in \mathbb{R}^{q \times p}$ where $\tilde{\phi}_{ih}^r = (\phi_{ih}^r - \bar{\phi}_i^r)/s_{\phi_i^r}^r$. ϕ_{ih}^r is the rank of ϕ_{ij} among elements of ϕ_i ($\{\phi_{ih}^r, h \in P^S\} = P^S$), $\bar{\phi}_i^r = (p+1)/2$ and $s_{\phi_i^r}^r = p(p+1)/12$ are the mean and the standard deviation of the ϕ_{ih}^r's repectively. The sample rank corre-

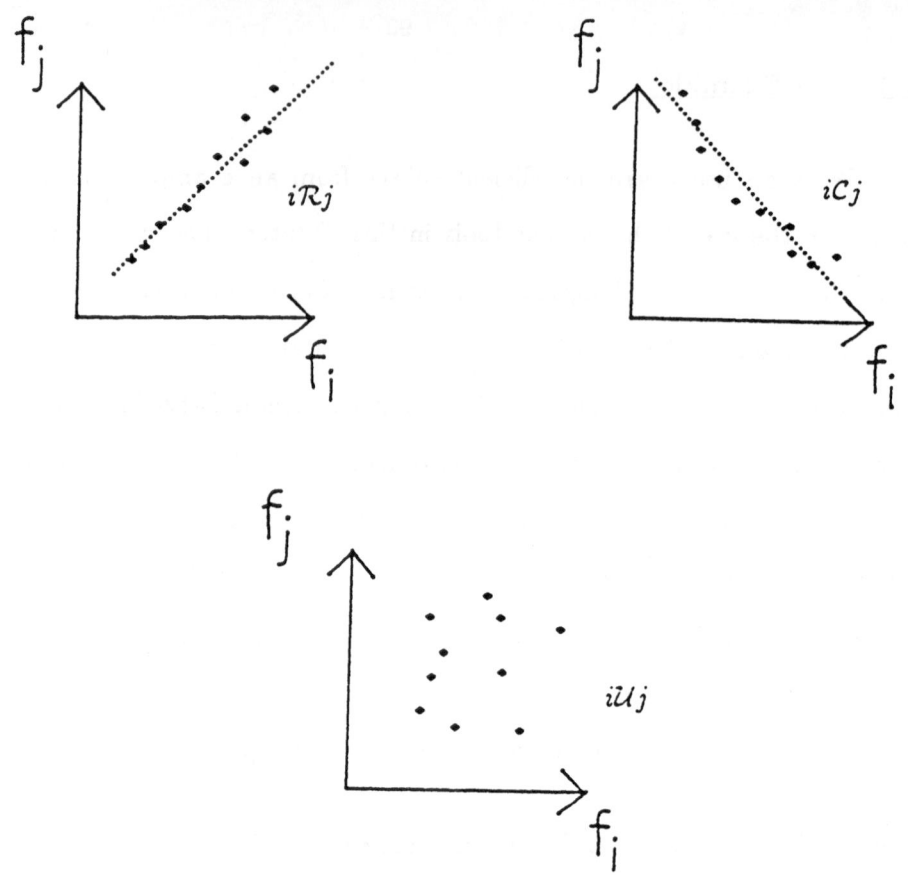

7.1 Performance Index Relations

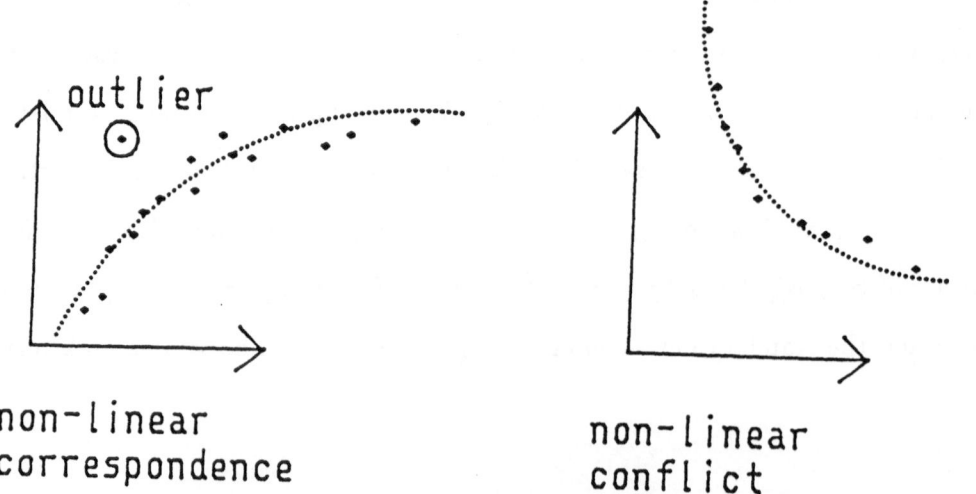

lation matrix will be given by

$$S = \tilde{M} \tilde{M}^T \qquad (7.2.1)$$

and $-1 \leq s_{ij} \leq 1$. The element s_{ij} is the sample rank correlation coefficient between performance indices i and j.

To define the three kinds of relations among performance indices, we may assign a threshold of correspondence $1 \geq s_r \geq 0$ and a threshold of conflict $0 > s_c \geq -1$ so that

$$\begin{cases} i\mathcal{R}j & \text{if } s_{ij} \geq s_r \\ i\mathcal{C}j & \text{if } s_{ij} \leq s_c \\ i\mathcal{U}j & \text{if } s_r > s_{ij} > s_c \end{cases} \qquad (7.2.2).$$

However, if the number of indices is large, it may be difficult for the designer to comprehend the relations between all possible index pairs. In such a case, we seek to cluster corresponding indices into a few index groups before we examine the relations between these groups instead of the individual indices. The following algorithm describes such a clustering procedure.

[STEP 1] Initialize a partition of the index set Q as

$$P^{(0)} = \left\{ Q_l^{(0)} = \{l\}, l \in M^{(0)} = \{1, 2, \ldots m^{(0)}\} \right\} . \qquad (A1.1)$$

where $m^{(0)} = q$. Let $k = 0$.

[STEP 2] Define the minimum linkage between any two index groups $Q_{l_1}^{(k)}, Q_{l_2}^{(k)} \in P^{(k)}$ as

$$L(l1, l2) = \arg \min \left\{ s_{i_1 i_2}, \quad i_1 \in Q_{l_1}^{(k)}, \quad i_2 \in Q_{l_2}^{(k)} \right\} \qquad (A1.2)$$

Identify two distinct index groups $Q_{l_1}\cdot, Q_{l_2}\cdot$ with maximum value of L. If $L(l1^*, l2^*) \leq s_r$, let $k^* = k$ and terminate. Otherwise, go to step 3.

[STEP 3] Construct a new partition as

$$P^{(k+1)} = \left\{ Q_l^{(k+1)}, \quad l \in M^{(k+1)} \right\}$$

$$\text{where} \quad Q_1^{(k+1)} = Q_{l_1\bullet}^{(k)} \cup Q_{l_2\bullet}^{(k)}$$

$$P^{(k+1)} - \{Q_1^{(k+1)}\} = P^{(k)} - \{Q_{l_1\bullet}^{(k)}, Q_{l_2\bullet}^{(k)}\} \tag{A1.3}$$

$$M^{(k+1)} = M^{(k)} - \{m^{(k)}\}$$

$$m^{(k+1)} = m^{(k)} - 1$$

Set $k = k + 1$. Go to step 2.

All partitions generated have the property that any two performance indices i, j in the same group will have $s_{ij} \geq s_r$, while the last one $P^{(k^*)}$ has the least number of groups. $P^{(k^*)}$ is determined by s_r and we denote it by $P(s_r)$. The same is applied to the $Q_l^{(k^*)}$'s, $M^{(k^*)}$ and $m^{(k^*)}$. The indices are clustered in a hierarchical manner: for any s_r such that $1 \geq s_r > 0$, $P(s_r)$ is among the partitions generated during execution of the algorithm for $s_r = 0$. The hierarchy of index groups can be represented graphically as in fig. 7.3.

Once a satisfactory partition is obtained which summarizes the correspondence relations, conflicting or uncorrelated relations are identified among groups. We define a **median linkage** between index groups Q_{l_1} and Q_{l_2} as

$$g_{l_1 l_2} \tag{7.2.3}$$

which is the median of $\{s_{ij}, i \in Q_{l_1}(s_r), j \in Q_{l_2}(s_r)\}$. Correspondingly, a median linkage matrix is

$$G(s_r) = \{g_{l_1 l_2}\} \in \Re^{m(s_r) \times m(s_r)} \tag{7.2.4}.$$

Therefore, half of all possible index pairs from groups $Q_{l_1}(s_r)$ and $Q_{l_2}(s_r)$ have their sample rank correlation less than $g_{l_1 l_2}$ (therefore conflicting to a certain degree). The structure of the index set can then be represented as $(P(s_r), G(s_r))$.

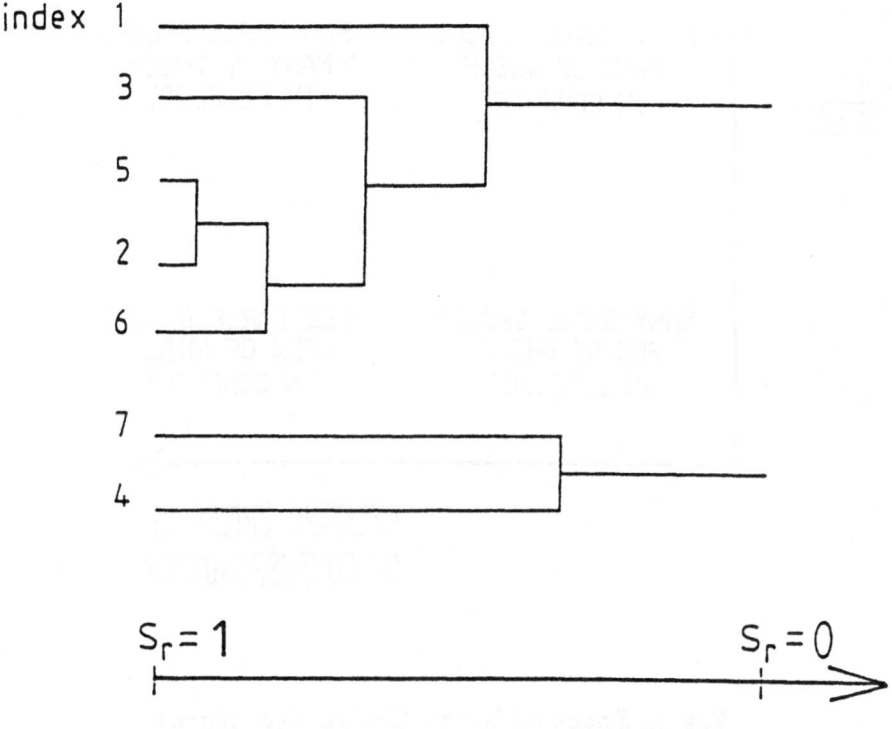

$S_r = 1$ $S_r = 0$

relaxing threshold of correspondence

7.3 A Hierarchy of Index Groups

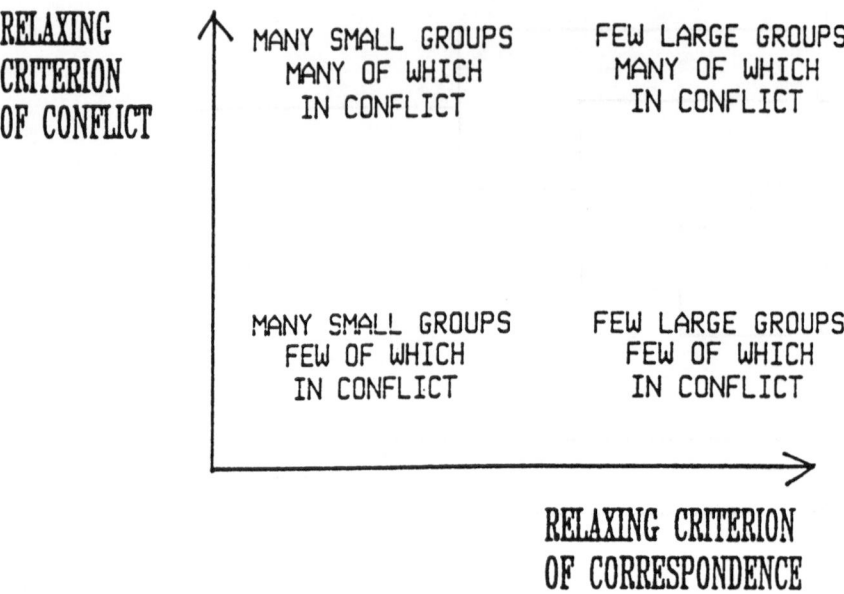

7.4 A Space of Index Group Structures

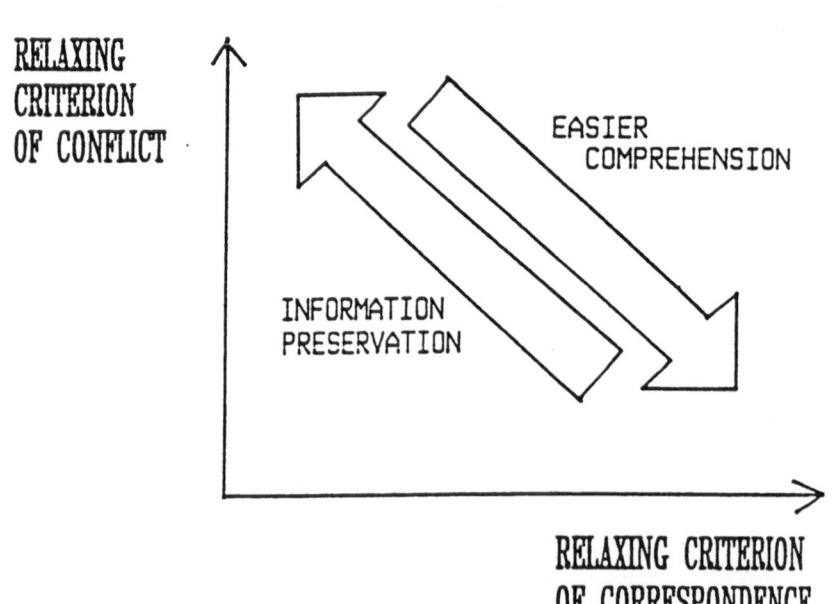

7.5 Trade-Off between Comprehension and Information Preservation

We can now generalize the definitions of relations (7.2.2) for index groups :

$$\begin{cases} Q_{l_1} \mathcal{R} Q_{l_2} & \text{if } g_{l_1 l_2} \geq s_r \\ Q_{l_1} \mathcal{C} Q_{l_2} & \text{if } g_{l_1 l_2} \leq s_c \\ Q_{l_1} \mathcal{U} Q_{l_2} & \text{if } s_r > g_{l_1 l_2} > s_c \end{cases} \qquad (7.2.5).$$

In this a way, the two threshold values define a structure space as in fig. 7.4. As illustrated in fig. 7.5, there is trade-off between information preservation and comprehension. The designer can be helped to explore this space by a display of the structure as in fig. 7.6. There is much scope for making such exploration highly interactive to enhance the designer's understanding.

7.2.1 The Example

We shall seek an index structure for the thirty-four indices of our example sample efficient subset.

The clustering algorithm was executed for $s_r = 0$ which gave 30 possible partitions, i.e. $P^{(0)}, P^{(1)}, P^{(2)}, \dots, P^{(29)}$. Therefore, if the maximum of the minimum linkages of index groups in $P^{(k)}$ is $L^{*(k)}$, assigning the threshold of correspondence s_r will give the partition

$$P(s_r) = P^{(k)} \quad \text{for which} \quad L^{*(k-1)} > s_r \geq L^{*(k)}. \qquad (7.2.1.1)$$

The number of groups can therefore be plotted against s_r as in fig. 7.7.

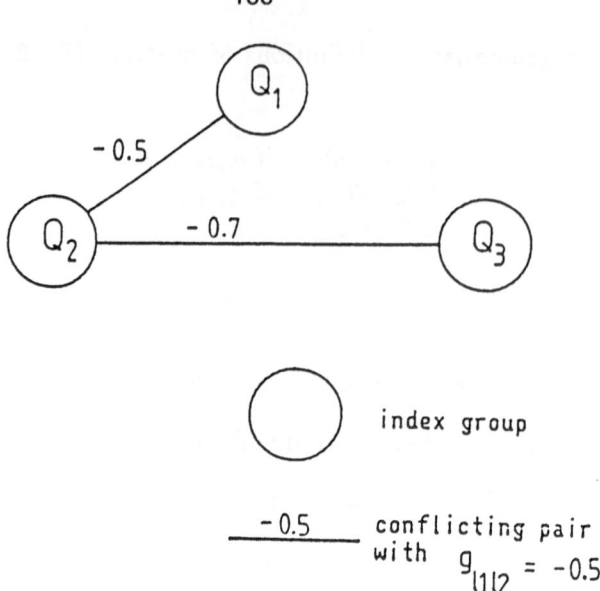

7.6 Graphical Display of an Index Group Structure

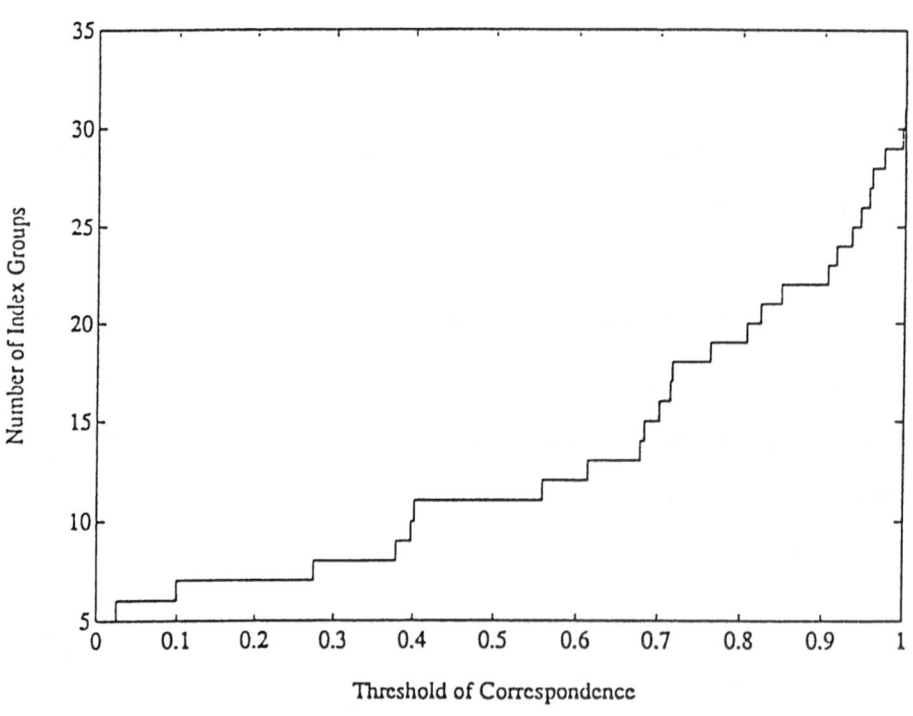

7.7 Variation of the Number of Index Groups   (Worked Example)

We chose $s_r = 0.2$ which gave $P(0.2) = P^{(27)}$ with 7 index groups :

$$Q_1(0.2) = \{1, 3, 8, 11, 16, 17, 21, 22, 34\}$$

$$Q_2(0.2) = \{2, 4, 7, 10, 15, 18, 23, 24\}$$

$$Q_3(0.2) = \{5\}$$

$$Q_4(0.2) = \{6, 9, 12, 13, 14, 25, 26, 31, 32, 33\} \qquad (7.2.1.2)$$

$$Q_5(0.2) = \{19, 20\}$$

$$Q_6(0.2) = \{27, 28\}$$

$$Q_7(0.2) = \{29, 30\}$$

This partition gave a median linkage matrix as

$$G(0.2) =$$

$$
\begin{pmatrix}
0.6918 & 0.0714 & -0.3585 & -0.6649 & -0.2129 & -0.4027 & 0.0556 \\
0.0714 & 0.6608 & 0.1394 & -0.0578 & -0.2522 & -0.5522 & -0.3101 \\
-0.3585 & 0.1394 & 1.0000 & 0.2833 & -0.1217 & 0.2210 & -0.6016 \\
-0.6649 & -0.0578 & 0.2833 & 0.6897 & 0.3034 & 0.2474 & 0.0847 \\
-0.2129 & -0.2522 & -0.1217 & 0.3034 & 0.8812 & 0.2307 & 0.1663 \\
-0.4027 & -0.5522 & 0.2210 & 0.2474 & 0.2307 & 0.8582 & -0.0479 \\
0.0556 & -0.3101 & -0.6016 & 0.0847 & 0.1663 & -0.0479 & 0.7010
\end{pmatrix}
$$

$$(7.2.1.3)$$

Assigning $s_c = -0.3$ gives 6 conflicting group pairs among the 21 possible pairs. The index structure can then be displayed graphically as in fig. 7.8. Also of interest is a display of sign entries of the sample rank correlation matrix with a permutation according to groups (fig. 7.9).

7.3 An Interactive Tool for Prescribing Index Bounds   [Ng 88]

When the designer has chosen a comprehensible structure (P, G) to represent the relations among indices, it can be used as a framework in which his wishes (expressed as index bounds) can be compared with the design possibilities (revealed in the sample efficient frontier Φ).

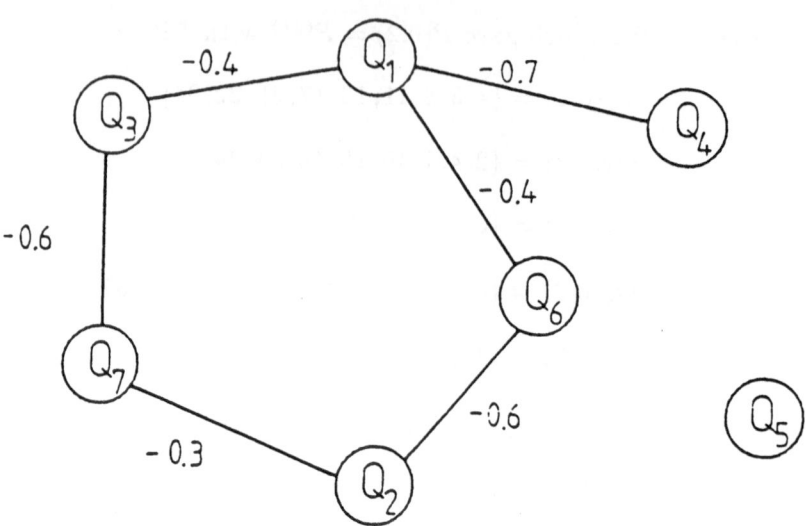

7.8 Graphical Display of the Index Group Structure (Worked Example)

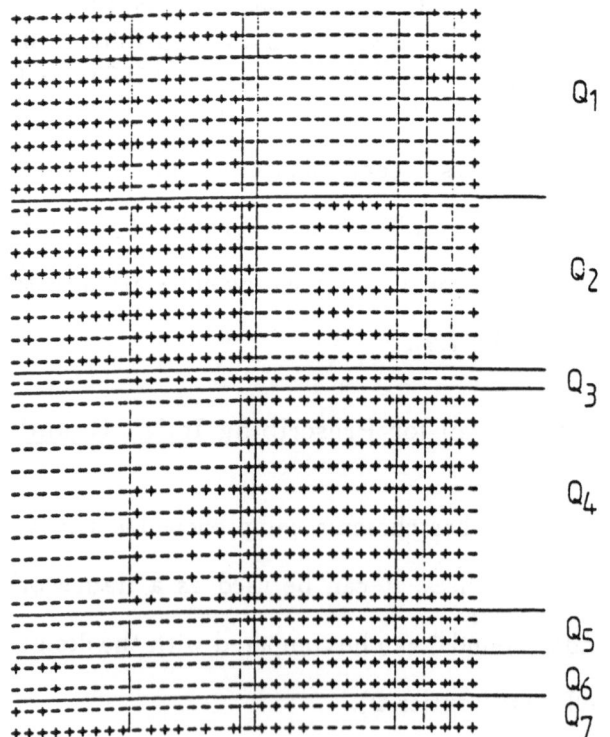

We have developed a novel interactive algorithm which accepts a vector of index bounds $c^{(0)} = (c_1^{(0)}, c_2^{(0)}, \ldots, c_q^{(0)})^T$ from the designer and reveals any stringent conflict in their attainment as well as the feasible trade-offs to resolve them. The approach is one of data visualization in which the designer is supported with graphical displays of data to make his trade-off decisions.

Before we describe the algorithm, we first define two essential types of graphs to be used :

[A2P1] A two-dimensional scatter plot of the points

$$\{(\phi_{ih}, \phi_{jh}), h \in P^S\} \cup \{(c_i, c_j)\}$$

as a projection of the sampled efficient frontier Φ and a vector of bounds c on the subspace of performance indices i, j.

[A2P2]

$$\left\{ \begin{array}{l} \dfrac{y_{hi} - \bar{y}_i}{n_i} \quad h \in P^S \\[2mm] \dfrac{c_i - \bar{y}_i}{n_i} \end{array} \right\} \qquad \text{verses} \quad [i]$$

where $\bar{y}_i$ and n_i are the median and mean absolute deviation of $\{y_{hi}, h \in P^S\}$ and $\mathcal{P}(\tilde{Q}) = ([1], [2], \ldots, [\tilde{q}])$ is a permutation of a subset of indices $\tilde{Q} \subset Q$.

The algorithm is described in the following. The designer supplies a vector of bounds $c^{(0)}$ as well as an index structure (P, G) $(P = \{Q_l, l \in M = \{1, 2, \ldots, m\}\}$ and $G = \{g_{l_1 l_2}\}$ as defined in (7.2.4)).

[STEP 1] Set $k = 1$.

[STEP 2] For each index group $Q_l \in P$, obtain $Q_l^{(k)} \subset Q_l \in P$ as a subset of performance indices whose current bounds $(c_i^{(k)})$ are satisfied by the least number of the sample efficient candidates. Select $i_l^{(k)} \in Q_l^{(k)}$ with the help of [A2P1] graphs for pairs of indices in $Q_l^{(k)}$, as the **active member** of the group. Let $\tilde{Q}^{(k)} = \{i_l^{(k)}, l \in M\}$ be the set of group active members.

[STEP 3] Obtain a permutation of the group active members as $\mathcal{P}(\tilde{Q}^{(k)})$. This permutation can be obtained either (i) from G, independent of $\tilde{Q}^{(k)}$, or (ii) from the sample rank correlation matrix of the $i_l^{(k)}$'s (i.e. $\tilde{S}^{(k)} = \{\tilde{s}_{l_1 l_2}\} \in \Re^{m \times m}$ where $\tilde{s}_{l_1 l_2} = s_{i_{l_1}^{(k)} i_{l_1}^{(k)}}$, or (iii) from the designer.

[STEP 4] Display an [A2P2]-type trade-off graph for $\tilde{Q}^{(k)}$ with permutation $\mathcal{P}(\tilde{Q}^{(k)})$ to help the designer appreciate the attainability of the current bounds as suggested by the sampled efficient frontier (fig. 7.10). If the bounds are acceptable, terminate. Otherwise, match design wishes with possibilities revealed in the trade-off graph by deciding new bound values for the displayed indices as $\{c_{i_{(k+1)}^{(k+1)}}^{(k+1)}, l \in M\}$ and retaining those of the others, i.e. $c_i^{(k+1)} = c_i^{(k)}, \forall i \in Q - \tilde{Q}^{(k)}$. Let $k = k + 1$. Go to STEP 2.

This algorithm is highly interactive and relies heavily on the designer's decision-making. There is no provable convergence and termination depends on the designer's satisfaction. The algorithm is essentially a structured dialogue between the designer and the computer in which conflicts among his initial set of prescribed index bounds are revealed in a progressive manner.

There are two criteria for a good permutation in STEP 3 : (i) most of the neighbours should be conflicting and (ii) the most serious conflicting pairs (index groups or active indices) as indicated by large negative values in G (7.2.4) or S (7.2.1), should be neighbouring to each other. A more detailed discussion is given in the appendix together with a simple heuristic procedure which gives a reasonable permutation based on a linkage matrix (either G or S).

7.3.1 The Example

We again use the problem example to illustrate the use of this algorithm.

Assume a vector of bounds was given as

$$
\begin{aligned}
c^{(0)} = (&1.00,\ 15.00,\ -0.23,\ 0.15,\ -0.30,\ 10.00,\ 10.00, \\
&15.00,\ 10.00,\ 10.00,\ 15.00,\ 10.00,\ 10.00,\ 15.00, \\
&15.00,\ -0.23,\ 0.20,\ 5.00,\ 0.20,\ 5.00,\ 0.10, \\
&3.00,\ 0.15,\ 0.15,\ 0.15,\ 0.15,\ 0.15,\ 0.15, \\
&0.10,\ 0.10,\ 0.10,\ 0.10,\ 0.10,\ 0.10)^T
\end{aligned}
$$

(7.3.1.1),

and the structure $(P(0.2), G(0.2))$ described in section 7.1.2 was used.

On the first iteration of the algorithm, we had

$$
\begin{aligned}
Q_1^{(1)} &= \{17\} \\
Q_2^{(1)} &= \{4, 24\} \\
Q_3^{(1)} &= \{5\} \\
Q_4^{(1)} &= \{31\} \\
Q_5^{(1)} &= \{19, 20\} \\
Q_6^{(1)} &= \{27\} \\
Q_7^{(1)} &= \{29, 30\}
\end{aligned}
$$

(7.3.1.2).

Only for groups 2, 5 and 7 were [A2P1]-type graphs needed for choosing the active members (fig. 7.10). The set of group active members was finally decided to be $\tilde{Q}^{(1)} = \{17, 24, 5, 31, 20, 27, 29\}$.

A permutation for the groups was derived from $G(0.2)$ using the heuristic algorithm in the appendix as $(5, 3, 7, 2, 6, 1, 4)$. Compared with the median linkage matrix $G(0.2)$ in (7.2.1.3), it can be seen that the two criteria for a good permutation as mentioned were both reasonably catered for. From this group permutation, we obtained $\mathcal{P}(\tilde{Q}^{(1)}) = (20, 5, 29, 24, 27, 17, 31)$. The trade-off graph for the group active indices was then plotted in fig. 7.11 with this permutation.

Based on fig. 7.11, it was decided that serious conflicts existed among the bounds for indices 27, 17 and 31 of group 6, 1 and 4 respectively. New bounds for them were set as $c_{27}^{(2)} = 0.30$, $c_{17}^{(2)} = 0.25$ and $c_{31}^{(2)} = 0.15$ while those of the others were retained.

At the second iteration ($k = 2$), the group active members were decided to be $\tilde{Q}^{(2)} = \{21, 4, 5, 25, 19, 27, 30\}$. The group permutation was again used and $\mathcal{P}(\tilde{Q}^{(2)}) = (19, 5, 30, 4, 27, 21, 25)$. New bounds were set as $c_{21}^{(3)} = 0.15$ and $c_{25}^{(3)} = 0.25$ while those of the others were retained. At the next iteration ($k = 3$), $\tilde{Q}^{(3)} = \{21, 4, 5, 33, 19, 27, 30\}$ and $\mathcal{P}(\tilde{Q}^{(2)}) = (19, 5, 30, 4, 27, 21, 33)$. One new bound was set as $c_{33}^{(4)} = 0.15$. On examining the $c_i^{(4)}$'s at the last iteration, it was decided that the atttainability is reasonable and the algorithm was terminated.

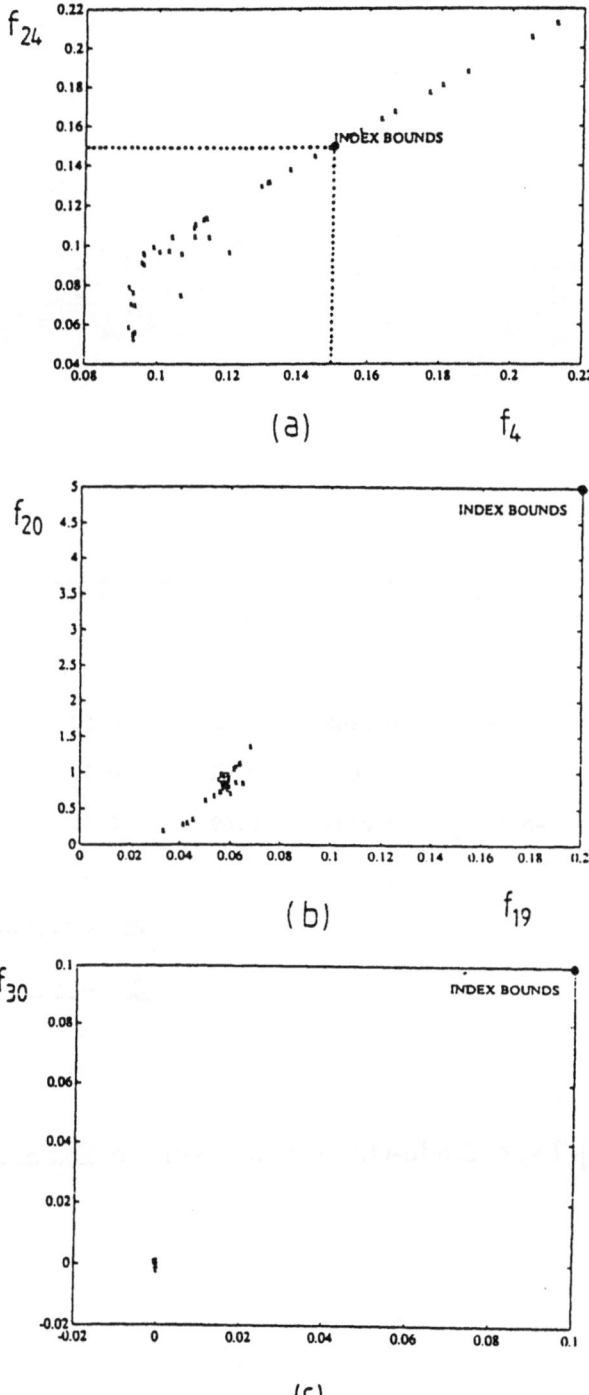

7.10 [A2P1]-Type Graphs for Choosing Group Active Members (Worked Example)

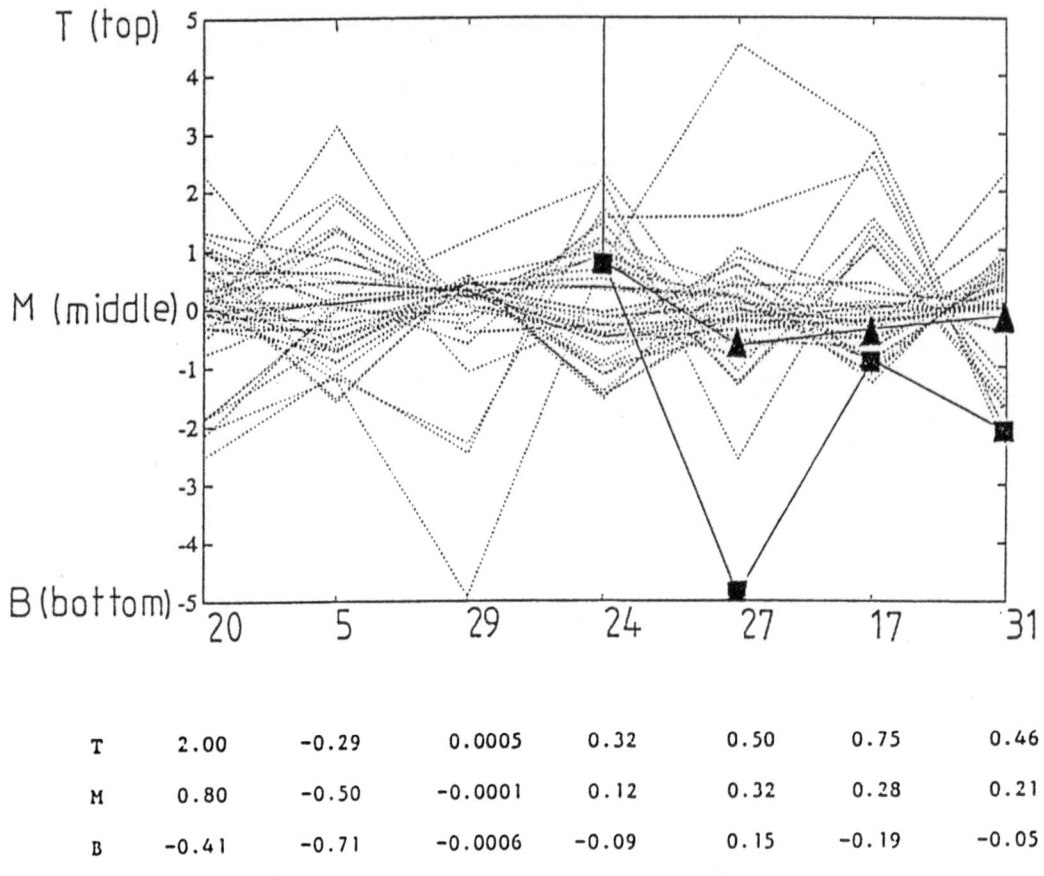

T	2.00	-0.29	0.0005	0.32	0.50	0.75	0.46
M	0.80	-0.50	-0.0001	0.12	0.32	0.28	0.21
B	-0.41	-0.71	-0.0006	-0.09	0.15	-0.19	-0.05

■ original bound

▲ relaxed bound

7.11 [A2P2]-Type Trade-Off Graph (Worked Example)

The final vector of compromise bounds is

$$c^{(0)} = (1.00, 15.00, -0.23, 0.15, -0.30, 10.00, 10.00,$$

$$15.00, 10.00, 10.00, 15.00, 10.00, 10.00, 15.00,$$

$$15.00, -0.23, 0.25, 5.00, 0.20, 5.00, 0.15, \qquad (7.3.1.3).$$

$$3.00, 0.15, 0.15, 0.25, 0.15, 0.30, 0.15,$$

$$0.10, 0.10, 0.15, 0.10, 0.15, 0.10)^T$$

The search stage was then executed to solve the resulting inequality formulation. An initial point was selected among the sample efficient set (with the evaluation tools to be described in the following chapter) which had favourable performance with respect to the modified bounds:

$$(0.87, 8.55, -0.67, 0.11, -0.49, 5.96, 0.51,$$

$$13.00, 1.27, 9.62, 1.52, 5.06, 0.79, 1.78,$$

$$8.55, -0.67, 0.27, 3.19, 0.06, 0.80, 0.14, \qquad (7.3.1.4),$$

$$1.30, 0.11, 0.07, 0.22, 0.04, 0.31, 0.04,$$

$$0.00, 0.00, 0.21, 0.04, 0.16, 0.02)^T,$$

violating 5 of the bounds (17, 21, 27, 31, 33). A final design obtained with the simplex polytope method on a dynamic minimax formulation (6.1.1.1) has a vector of performance indices as

$$(0.99, 6.92, -0.52, 0.10, -0.51, 7.98, 0.11,$$

$$14.80, 1.29, 8.56, 0.90, 5.17, 0.75, 6.32,$$

$$6.92, -0.52, 0.25, 2.91, 0.09, 1.00, 0.14, \qquad (7.3.1.5)$$

$$0.98, 0.10, 0.15, 0.17, 0.04, 0.23, 0.03,$$

$$0.00, 0.00, 0.16, 0.03, 0.13, 0.02)^T$$

which violated one bound (31) slightly.

CHAPTER 8

EVALUATION TOOLS

It is not difficult to appreciate the central role that the sample efficient subset plays in the design strategy. It captures the design history and reveals design possibilities, with which the designer's wishes are matched using the formulation tools described in the previous chapter.

As design proceeds, the subset is continually updated. Trajectories of new candidate designs are generated during iterations of the search stage as well as the exploratory stage. Alternative initial designs and parametrizations from level I give rise to trajectories in design parameter spaces of different dimensions and mappings to the performance index space. The subset's structure is illustrated in fig. 8.1. There is a need to maintain the design data for a correct representation of the subset.

Apart from evaluations conducted on individual candidates, a proper evaluation of the design history is essential. The designer should seek to gain a holistic view of the design process as well as a proper context in which new candidates are evaluated. However, such evaluation is an unstructured task which may be conducted in a variety of different ways depending on the designer's need and preference. Minimal structure should be imposed on the designer-computer interaction. *A language for evaluation is required.* As the design history is captured by the sample efficient subset, the evaluation language should have a flexible data interrogation facility.

We therefore propose a database view of the sample efficient subset to support evaluations with database queries. Unlike the last two chapters in which the chief

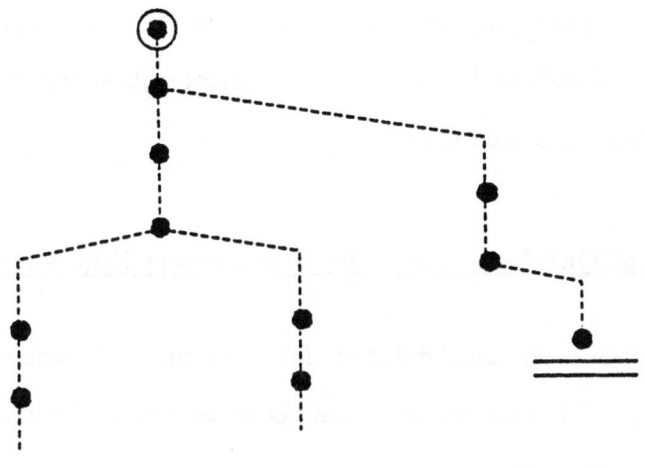

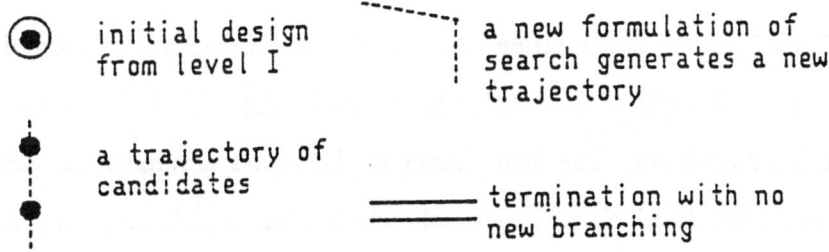

8.1 Structure of a Sample Efficient Subset

effort was spent in developing procedures for relatively simple data structures, it is the design of a database for the sample efficient subset which dominates the development of the evaluation tools.

8.1 A Relational Database Model for the Sample Efficient Subset

We have chosen to design a database for the sample efficient subset using the relational model [Cod 71] because of its simplicity as well as the natural correspondence with our design data.

A relational database model consists of **relations** which are sets of related fields. Each instance of a relation consists of a set of **tuples**, or **records**. Each relation can be thought of as a table. Each row in the table is a tuple while each column is an **attribute**. The field entries in the tuples assume their values from the corresponding **field domains**. Each relation has a **primary key** as one or more fields which uniquely identifies any tuple within a table. The database itself is then represented by a data dictionary which is again a relation, between relation name, number of tuples, field name, field domain and field reference. **Integrity constraints** may be imposed as dependencies among tuples which must be satisfied by any instance of the database.

To maintain the subset as a relational database, the various design parameter spaces in which trajectories reside are represented by different relations, which we shall name as X-relations (fig. 8.2a). Each tuple represents a candidate in the space. The primary key is N# which is an integer-valued identifier of the candidate.

Similarly, the performance index space is represented as a Y-relation (fig. 8.2b). Each tuple represents a vector of performance indices as the mapping of a candidate

```
| N# | X_1 | X_2 | ... | X_n |
|    |     |     |     |     |
|————|—————|—————|—————|—————|
|    |     |     |     |     |
|    |     |     |     |     |
```

N# — 1,2,3,...   — identifiers of candidate
X_1,X_2,...,X_n — design parameters

(a)   X—Relation

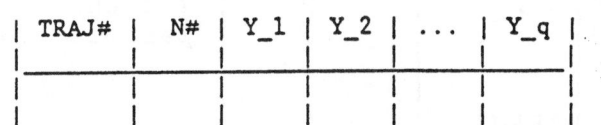

```
| TRAJ# | N# | Y_1 | Y_2 | ... | Y_q |
|       |    |     |     |     |     |
|———————|————|—————|—————|—————|—————|
|       |    |     |     |     |     |
|       |    |     |     |     |     |
```

TRAP# — 1,2,3,...   — identifiers of trajectory
Y_1,Y_2,...,Y_q   — performance indices

(b)   Y—Relation

```
| TRAJ# | PTRAJ | XON | FORM | PARA |
|       |       |     |      |      |
|———————|———————|—————|——————|——————|
|       |       |     |      |      |
|       |       |     |      |      |
```

PTRAP — identifiers of preceding trajectory
XON   — identifiers of initial point on PTRAJ
FORM — 1,2,3,...   — formulation of the numerical search
PARA — name of the X—relation representing the parameter
        space in which the trajectory resides

(c)   T—Relation

8.2 Relational Database Model of a Sample Efficient Subset

RELATION	NTUPLES	FIELD	DOMAIN	REFERENCE
X1	nx1	N#	Z	
X1	nx1	X_1	R	design parameters
X1	nx1	X_2	R	for the first
:	:	:	:	parametrization
:	:	:	:	
X2	nx2	N#	Z	
X2	nx2	X_1	R	design parameters
X2	nx2	X_2	R	for the second
:	:	:	:	parametrization
:	:	:	:	
Y	ny	TRAJ#	Z	
Y	ny	N#	Z	
Y	ny	Y_1	Z	
Y	ny	Y_2	Z	performance indices
:	:	:	:	
:	:	:	:	
Y	ny	Y_q	Z	
T	nt	TRAJ#	Z	
T	nt	PTRAJ	Z	
T	nt	XON	Z	
T	nt	FORM	Z	
T	nt	PARA	string	

8.3 Data Dictionary of the Relational Database Model

in some design parameter space. The primary key is TRAJ#, an integral identifier for trajectories together with N#, the candidate's identifier.

A trajectory is generated by a numerical search with an initial point which may belong to another trajectory. This chronological dependency is represented by yet another relation, the T-relation (fig 8.2c). Each tuple represents a trajectory. TRAJ# alone is the primary key. PTRAJ and XON may take null values if the initial point is not obtained from another trajectory. FORM is an identifier for the formulation of the search which results in the trajectory. PARA is the name of the X-relation which represents the parameter space in which the trajectory resides.

The resulting data dictionary is illustrated in fig 8.3. To ensure that all entries are not disproved of efficiency (when the performance indices are dominated by those of some other entries), an integrity constraint of **internal efficiency** is imposed as: in any instance of the Y-relation, no tuple has its performance index attributes dominated (as in (3.3.1)) by those of any other tuple.

8.2 Data Manipulation Procedures

Standard data manipulation procedures for databases can be divided into the following categories: (1) dictionary, (2) creation (of new tuples), (3) deletion (of tuples), (4) update (of tuples), (5) query (retrieval), (6) views, (7) protection, (8) sharing, (9) recovery and (10) query optimization. However, only the first five categories will be useful to the designer in maintaining and interrogating the database for evaluation. The last five are concerned with security and implementation.

8.2.1 A Database Query Language

We shall describe a popular query language and illustrate how it can be used in evaluations. Assume a design project was conducted using the design strategy where level I was iterated three times for three alternative feasible sets with different parametrizations. The sample efficient subset is maintained as a relational database which consists of T-TURBO as the T-relation, Y-TURBO as the Y-relation and three X-relations as PARA_1, PARA_2 and PARA_3 (fig. 8.4).

SEQUEL (Structured English QUEry Language, also known as SQL) [Cha 74] is a database query language developed for System R, a relational database management system designed by IBM. The basic syntax of queries is a mapping operation

$$
\begin{array}{ll}
\text{SELECT} & \textit{(expression(s))} \\
\text{FROM} & \textit{(relation(s))} \\
\text{WHERE} & \textit{(predicate)}. \qquad (8.2.1.1)
\end{array}
$$

The WHERE-clause defines a **horizontal subsetting** to extract a collection of tuples, followed by a **vertical subsetting** defined by the SELECT-clause to extract a subset of attributes. The expressions in the SELECT-clause are arithmetic, all of which are either in field entries such as $Y_1 + Y_2$, or in **aggregate functions** of attributes such as $MAX(Y_1) / MAX(Y_2)$, which returns a single argument. The predicate statement in the WHERE-clause may also comprise such expressions. Expressions and relations are separated by commas in the corresponding clause. A simple example of a query is

$$
\begin{array}{ll}
\text{SELECT} & \text{TRAJ\#} \\
\text{FROM} & \text{T-TURBO} \\
\text{WHERE} & \text{PARA} = \text{PARA\_1} \qquad (8.2.1.2)
\end{array}
$$

which returns all the trajectories in the design project generated in the parameter space PARA_2.

N#	X1_1	X1_2	X1_3

PARA_1

N#	X2_1	X2_2	X2_3

PARA_2

N#	X3_1	X3_2	X3_3

PARA_3

TRAJ#	N#	Y_1	Y_2	...	Y_n

Y—TURBO

TRAJ#	PTRAJ	XON	FORM	PARA

T—TURBO

8.4 Relational Database Model (An Example)

Mappings can also be nested. For instance,

SELECT	*
FROM	PARA_1
WHERE	N# is_in

	SELECT	N#
	FROM	Y-TURBO
	WHERE	$Y_1 \leq 0$
	AND	TRAJ# is_in

		SELECT	TRAJ#
		FROM	T-TURBO
		WHERE	PARA = PARA_1 (8.2.1.3)

returns all field entries of the candidates of a particular parametrization (PARA_1) with the first performance index satisfying $f(Y_1) \leq 0$.

Partitioning of a table into **tuple groups** is possible with the "GROUP_BY"-operator. For instance,

SELECT	TRAJ#
FROM	Y-TURBO
GROUP_BY	TRAJ#
HAVING	count(N#) $\geq$ 10 (8.2.1.4)

returns trajectories having more than ten candidates. The HAVING-clause is the equivalent of WHERE-clause for tuple groups.

8.2.2 Maintenance of the Design Database

The chief concern in maintaining the database is the integrity constraint of internal efficiency. This can be ensured by an automatic deletion of dominated candidate after each creation or update operation. Suppose a new trajectory is obtained which creates new tuples in the relations Y-TURBO, T-TURBO and one of the X-relations, the candidates to be deleted are identified as

SELECT	TRAJ#, N#	
FROM	Y-TURBO T	
WHERE	<TRAJ#, N#> is_in	
	(SELECT	TRAJ#, N#
	FROM	Y-TURBO
	WHERE	TRAJ# not_in T.TRAJ#
	AND	N# not_in T.N#
	AND	Y_1 $\leq$ T.Y_1
	AND	Y_2 $\leq$ T.Y_2
	⋮	⋮
	AND	Y_q $\leq$ T.Y_q).

$$(8.2.2.1)$$

where T is a synonym of Y-TURBO.

8.3 An Evaluation Language

An advantage of the relational database is the uniformity of the data representation. In fact, a basic measure of the selective power of a database query language is relational completeness [Cod 72] which is based on the relational calculus [Dat 80]. The query language SEQUEL we described is indeed relationally complete. However, a query language only performs the retrieval of data. If considerable transformations are required on the data, a host language interface is often used in which the queries are embedded. The host language is often a full power high level language. In the case of SEQUEL, the host language is PL/I.

Evaluation procedures on a large number of candidates are essentially comparisons. To formulate evaluation queries for a relational database which represents a sample efficient subset, tuples are compared with each other, as well as with specifications. Such comparisons are often conducted on the entries of the fields Y_1, Y_2, ..., Y_q, which represent the performance indices, using certain criteria of merit. According to such criteria, tuples in a Y-relation may be labelled as good designs or field entries labelled good index values, while tuples in a T-relation are labelled as good trajectories.

Although such operations may induce considerable transformation, the structure of computations is sufficiently consistent that it suffices to employ other data manipulation procedures without calling for a host language. Creation and update procedures can be used in conjunction with queries so that temporary relations may be created to store intermediate outputs of queries, updated according to required computations, and subsequently queried for the results. For instance, to find the candidate(s) which satisfy $Y\_1 \leq 0$ as well as the largest number of a set of bounds for other indices $c_i, i = 2, 3, \ldots, q$, first create a temporary relation TEMP as

ASSIGN TO TEMP

SELECT   *

FROM Y-TURBO

WHERE $Y\_1 \leq 0,$ (8.3.1a)

expand a new column for TEMP

EXPAND TABLE TEMP

    ADD COLUMN NBOUNDS(INTEGER),   (8.3.1b)

compute the values of NBOUNDS by a series of updates as

UPDATE TEMP

SET $\qquad$ $Y\_i = 1$

WHERE $\qquad$ $Y\_i \leq c_i$

UPDATE TEMP

SET $\qquad$ $Y\_i = 0$

WHERE $\qquad$ $Y\_i > c_i$ $\hfill$ (8.3.1c)

for $i = 2, 3, \ldots, q$ and

UPDATE TEMP

SET $\qquad$ NBOUNDS $= Y\_1 + Y\_2 + \ldots + Y\_q,$ $\hfill$ (8.3.1d)

and finally, execute the query

SELECT $\qquad$ TRAJ#, N#, NBOUNDS

FROM $\qquad$ TEMP

WHERE $\qquad$ NBOUNDS =

SELECT MAX(NBOUNDS)

FROM $\qquad$ TEMP. $\hfill$ (8.3.1e)

(The data manipulation procedures are according to the syntax of SEQUEL-2 [Cha 76], an expanded version of SEQUEL as a full scale data manipulation language.)

However, it is possible to enrich the syntax of the query language SEQUEL itself to achieve such compound queries for comparison in a single statement. We propose the following two extensions to SEQUEL :

(1) the restriction to arithmetic expressions should be relaxed to allow boolean elements to be mixed with arithmetic ones;

(2) parametrized macro expressions of field values should be allowed.

We shall call the resulting facility an evaluation language. Allowing boolean elements enables field entries to be labelled and counted, and tables with boolean entries generated, while a set of parametrized macros may be maintained as different criteria of merit.

With the evaluation language, the compound query in (8.3.1) can be issued with a single statement

SELECT	TRAJ#, N#, NFIELDS
FROM	Y-TURBO
WHERE	$Y\_1 \leq 0$
AND	NFIELDS =
	SELECT MAX(NFIELDS)
	FROM Y-TURBO. (8.3.2)

The macro NFIELDS is defined as $A_1 + A_2 + \ldots + A_q$ where $A_i = (Y\_i \leq c_i)$, $i = 2, 3, \ldots, q$ are boolean expressions. The macro is parametrized by the bounds c_i's. (Note that applying the NFIELDS macro to a tuple (row) with the boolean expressions A_i's is similar to repeatedly applying the SEQUEL aggregate function COUNT to an attribute (column) of a table subsetted with different predicates.)

To find out the number of candidates which satisfy each of the bounds,

SELECT	SUM(A_2), SUM(A_3), ..., SUM(A_q)
FROM	Y-TURBO
WHERE	$Y\_1 \leq 0$
AND	NFIELDS =
	SELECT MAX(NFIELDS)
	FROM Y-TURBO (8.3.3)

To compare two trajectories, TRAJ# $=1$, 2 say, by finding out the number of candidates which satisfy a set of bounds $c_i, i = 1, 2, \ldots, q,$

SELECT	SUM((TRAJ#=1)), SUM((TRAJ#=2))
FROM	Y-TURBO
WHERE	TRAJ# in $\{1, 2\}$
AND	A_1
AND	A_2
$\vdots$	$\vdots$
AND	A_q

$$(8.3.4)$$

8.4 Discussion

While the evaluation language provides an interaction medium for the designer to articulate his queries for evaluation, the results of queries are best presented in graphical forms. In a similar application of database query language in supporting managerial decision making, Sprague and Carlson [Spr 82] argued that as the scope of queries is usually derived from the data presented as results of another function (possibly another query), it is desirable to have a graphical interface for querying. The user may then define the scope of a query by pointing or circling regions in a scattered plot of points which represent tuples generated as results of the last query.

In our Pro-Matlab prototype, we have implemented neither the evaluation language nor a true relational database as the effort involved is not justified. As an alternative, the relations of the proposed database are stored as matrices which are maintained manually. An adequate set of evaluation queries was implemented as matrix functions which make use of the "find" command, the relational operations, together with the submatrix addressing capabilities of Pro-Matlab.

CHAPTER 9

TWO DESIGN EXAMPLES

We describe two examples in the design of linear time-invariant (LTI) multi-variable controllers to illustrate the use of the design strategy. The first example is relatively straightforward and serves to exemplify the resulting design process in a typical application of the strategy. The second example is a more involved problem in which the strategy was used in a most flexible way. It affirms that the strategy should assume a supporting role to the designer without imposing undue restrictions on necessary situational actions [Suc 87].

In both examples, hard constraints were not explicitly formulated but were included in the set of performance indices instead. This was so that auxiliary optimizations were simpler unconstrained problems. In fact, true hard constraints are not common as avoiding the vicinity of strict bounds was always considered desirable for safety, among other reasons.

9.1 A Nuclear Powered Turbo-Generator

The plant to be controlled is a 1072 MVA nuclear powered turbo-generator [Lim 79]. The plant is inherently non-linear and a mathematical model was derived from physical considerations. For design purposes, the model was linearized about a full-load operating point which results in a linear state-space model with two inputs (the throttle valve postion Y_g and the excitation control U_e), ten states and two outputs (the generator terminal voltage V_t and the generator load angle δ) (Appendix B).

9.1.1 Level I: Innovation - Feasible Set Generation

The aim of this level is to obtain an initial design *as an element of a parametrized feasible set.* The design of an LTI controller for the plant was reported in [Lim 85] whose objective was to obtain a fast non-interacting transient response. To illustrate the design strategy, it sufficed to adopt the reported design procedures. However, we had to seek a parametrization of the initial controller to construct a feasible set. The design specifications were to be elaborated in level II.

The configuration of the controller designed is shown in fig. 9.1. The controller was designed using multivariable frequency domain design methods [Mac 77], [Pos 81]. The full description of such a controller was given which is composed of eight sub-controllers (fig. 9.2). We have chosen to obtain a parametrization based on this decomposition. Altogether fifteen parameters were identified as $x_i, i = 1, 2, \ldots, 15$ (fig. 9.2). Some of these were scaled versions of the parameters used in [Lim 85]. We used the reported controller as the initial design for level II.

The initial design obtained was unsatisfactory in the response of the generator load angle δ as shown in fig. 9.3 (the final design in [Lim 85] was actually a modified version with extra dynamic elements which were not described).

9.1.2 Level II: Trade-offs - Matching Wishes and Possibilities

Our design objectives included closed loop stability, dynamic performance (tracking and decoupling), structural stability, control effort and noise rejection. However, close loop stability and performance were the primary objectives. Since the control effort and the noise rejection of the initial design were considered satis-

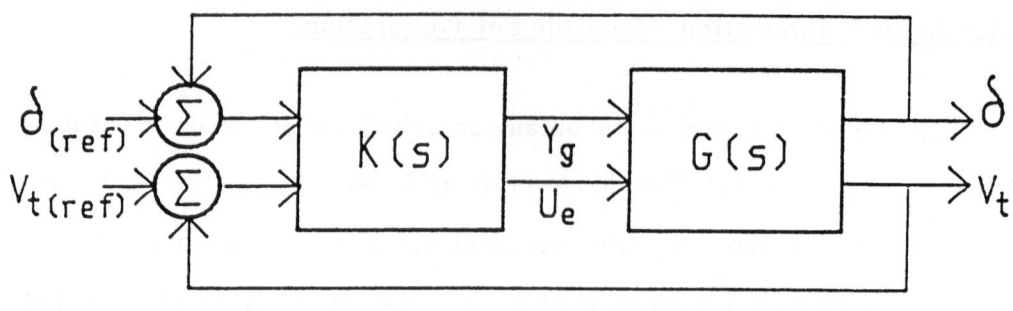

9.1 Controller Configuration (NPTG Example)

$$K(s) = \underbrace{K_1 K_2}_{\substack{\text{low frequency} \\ \text{sub} - \text{controllers}}} \quad \underbrace{K_3 K_4(s) K_5}_{\substack{\text{medium frequency} \\ \text{sub} - \text{controllers}}} \quad \underbrace{K_6 K_7(s) K_8}_{\substack{\text{high frequency} \\ \text{sub} - \text{controllers}}}$$

$K_1 =$ a real approximation of $[G(jw_h)]^{-1}$

(obtained by the ALIGN algorithm [Mac 77])

$$K_2 = \begin{pmatrix} k_1 & 0 \\ 0 & k_2 \end{pmatrix}$$

$K_3 =$ a real approximation of the eigenframe of

$\quad G(jw_m) K_1 K_2$

$$K_4 = \begin{pmatrix} k_3(s) & 0 \\ 0 & k_4(s) \end{pmatrix} \quad \text{where}$$

$$k_3(s) = \frac{s^2 + p_1 s + q_1}{s^2 + p_2 s + q_2} \frac{k_5(s + w_1/\alpha_1)}{s + w_1}.$$

$$k_4(s) = \frac{k_6(s + \omega_2/\alpha_2)}{s + \omega_2}$$

$K_5 = K_3^{-1}$

$K_6 =$ a real approximation of the right singular frame of

$\quad M = G(0) K_1 K_2 K_3 K_4(0) K_5$

$$K_7(s) = \begin{pmatrix} k_5(s) & 0 \\ 0 & k_6(s) \end{pmatrix} \quad \text{where}$$

$$k_5(s) = 1 + \frac{\alpha_3 \gamma_2}{s}$$

$$k_6(s) = 1 + \frac{\alpha_3 \gamma_1}{s}$$

$\quad \gamma_1 \geq \gamma_2$ are the two singular values of M

$K_8 =$ a real approximation of the inverse of the left singular frame of M

$$\begin{aligned}
\omega_h &= \exp(x_1) \\
k_1 &= -\exp(x_2) \\
k_2 &= -\exp(x_3) \\
\omega_m &= \exp(x_4) \\
p_1 &= x_5 \\
q_1 &= x_6^2 \\
p_2 &= x_7 \\
q_2 &= x_8^2 \\
\omega_1 &= \exp(x_9) \\
k_5 &= \exp(x_{10}) \\
\alpha_1 &= x_{11} \\
\omega_2 &= \exp(x_{12}) \\
k_6 &= \exp(x_{13}) \\
\alpha_2 &= x_{14} \\
\alpha_3 &= x_{15}
\end{aligned}$$

9.2 Controller Parametrization (NPTG Example)

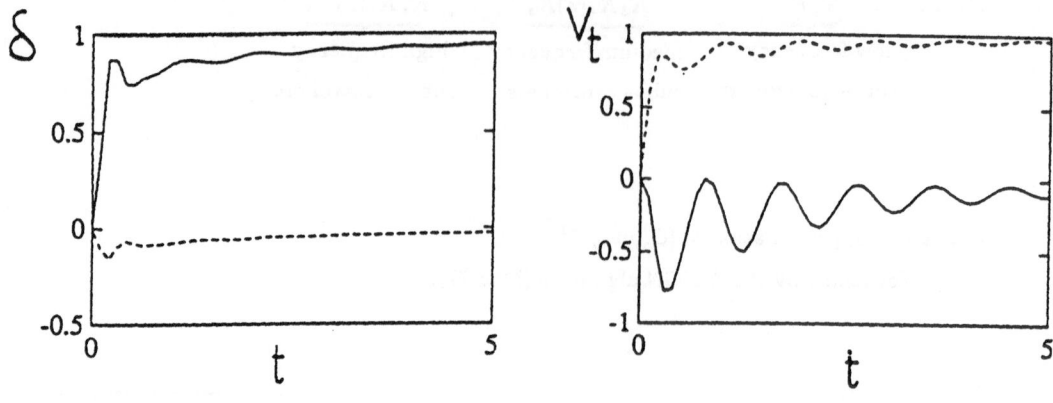

9.3 Close Loop Step Responses of the Initial Design (NPTG Example)

factory, they were not dealt with in the exploratory stage.

(1) Exploratory Stage

Four aggregate performance indices were formulated as

$$\hat{f}_{\hat{i}} : \Re^{15} \rightarrow \Re, \quad \hat{i} \in \hat{Q} = \{1, 2, 3, 4\}, \tag{9.1.2.1}$$

the details of which are described in fig. 9.4. Each index was to be minimized for good level of the indicated objectives. The initial design achieved $\hat{\mathbf{f}} = (-0.17, 320, -0.076, 0.76)^T$.

The numerical search tools described in section 6.1 were used. Design iterations were executed as the prescription of good and bad values for the $\hat{i}$'s and the optimization of the minimax formulation (4.3.4) using the simplex polytope method, whose terminations were at the designer's discretion.

For illustration purpose, only two iterations were executed. We describe them in fig. 9.5. The first iteration aimed to decrease dynamic coupling (which was a success) while the second aimed to improve damping (which might be considered a failure). The two iterations produced 19 distinct controllers whose aggregate performance index values are shown in fig. 9.6. It can be observed that the last nine are very similar. (For a full scale design exercise, more effort may have to be spent in this stage to explore more about the aggregate indices and to generate more candidate designs.)

An elaborate set of performance indices were then chosen to represent all design objectives. As a result, thirteen indices were formulated as

$$f_i : \Re^{15} \rightarrow \Re, \quad i \in Q = \{1, 2, \ldots, 13\}, \tag{9.1.2.2}$$

i	DEFINITION	INDICATES
1	maximum real part of the close loop poles	control loop stability
2	maximum modulus of the close loop poles	structural stability
3	negative of the minimum damping of the close loop poles	damping
4	maximum absolute value of the step (in δ) response of V_i	dynamic coupling

9.4 Aggregate Performance Indices (NPTG Example)

	Designer prescribes				Candidate design achieves			
	$\hat{f}_1^{g/b}$	$\hat{f}_2^{g/b}$	$\hat{f}_3^{g/b}$	$\hat{f}_4^{g/b}$	$\hat{f}_1$	$\hat{f}_2$	$\hat{f}_3$	$\hat{f}_4$
					-0.170	315.3	-0.076	0.761
good	-0.155	315.0	-0.077	0.100				
bad	-0.150	320.0	-0.050	0.770	-0.226	315.2	-0.094	0.131
good	-0.210	315.0	-0.200	0.130				
bad	-0.200	320.0	-0.094	0.150	-0.200	315.2	-0.095	0.146

9.5 Design Iterations in the Exploratory Stage (NPTG Example)

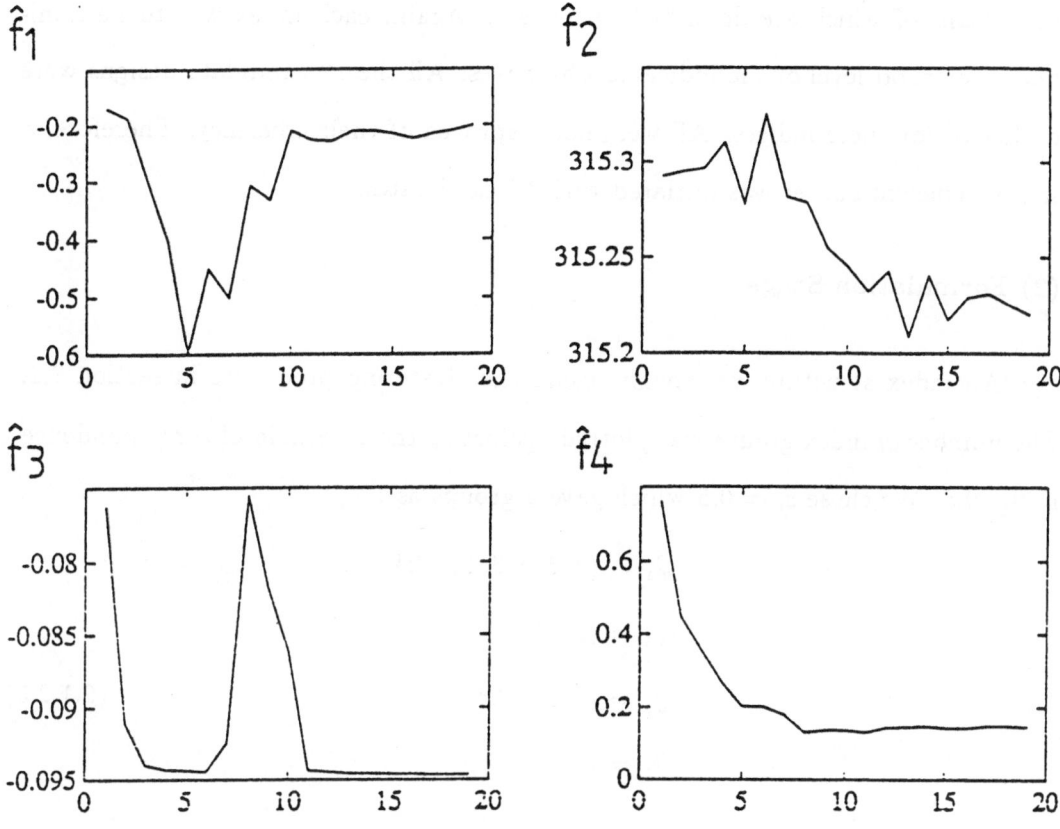

9.6 Aggregate Performance Indices' Values (NPTG Example)

the details of which are described in fig. 9.7. Again, each index was to be minimized for good level of the indicated objectives. All the 19 candidate designs were
evaluated for these indices. All were not disproved of their efficiency. Therefore, a
sample efficient subset was initiated with 19 candidates.

(2) Formulation Stage

An index structure was sought using the clustering procedure in section 7.2.
The number of index groups was plotted against s_r the threshold of correspondence
in fig. 9.8. We chose $s_r = 0.5$ which gave 5 groups as :

$$Q_1 = \{3, 5, 6, 9, 11, 12\}$$

$$Q_2 = \{2, 8\}$$

$$Q_3 = \{7, 10, 13\} \tag{9.1.2.3}$$

$$Q_4 = \{1\}$$

$$Q_5 = \{4\}$$

This partition gave a median linkage matrix as

$$G = \{g_{l_1 l_2}\}$$

$$= \begin{pmatrix} 0.8956 & 0.3798 & -0.7693 & 0.3333 & -0.2333 \\ 0.3798 & 0.9974 & -0.3921 & -0.1509 & -0.1711 \\ -0.7693 & -0.3921 & 0.7860 & -0.2035 & 0.1825 \\ 0.3333 & -0.1509 & -0.2035 & 1.0000 & 0.4614 \\ -0.2333 & -0.1711 & 0.1825 & 0.4614 & 1.0000 \end{pmatrix} \tag{9.1.2.4}$$

Assigning the threshold of conflict to its minimum possible value, i.e. $s_c = 0$ gave
6 conflicting pairs among the 20 possible pairs. The index structure is graphically
displayed in fig. 9.9.

A physical interpretation may be as follows. All indices in Q_1 apart from index
3 are concerned with the performance of V_t while those in Q_3 are concerned with that

i	DEFINITION	INDICATES
1	maximum real part of the close loop poles	control loop stability
2	maximum absolute step response of Y_g	control effort
3	maximum absolute step response of U_e	control effort
4	IAE of step (in V_t) response of V_t	command tracking
5	IAE of step (in δ) response of δ	command tracking
6	IAE of step (in V_t) response of δ	coupling
7	IAE of step (in δ) response of V_t	coupling
8	-40dB bandwidth of CLTF of V_t	noise rejection
9	-40dB bandwidth of CLTF of δ	noise rejection
10	negative of minimum frequency domain gain (up to 10 rad/s) from V_t to V_t	harmonic tracking
11	negative of minimum frequency domain gain (up to 10 rad/s) from V_t to V_t	harmonic tracking
12	maximum frequency domain gain (up to 10 rad/s) from V_t to δ	harmonic coupling
13	maximum frequency domain gain (up to 10 rad/s) from δ to V_t	harmonic coupling

9.7 Performance Indices (NPTG Example)

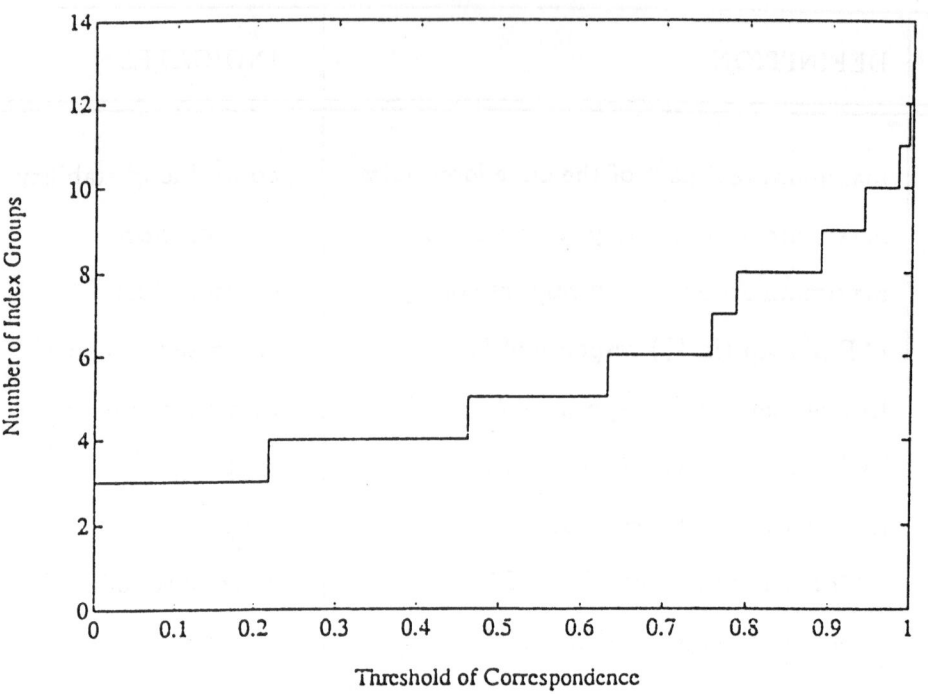

9.8 Variation of the Number of Index Groups (NPTG Example)

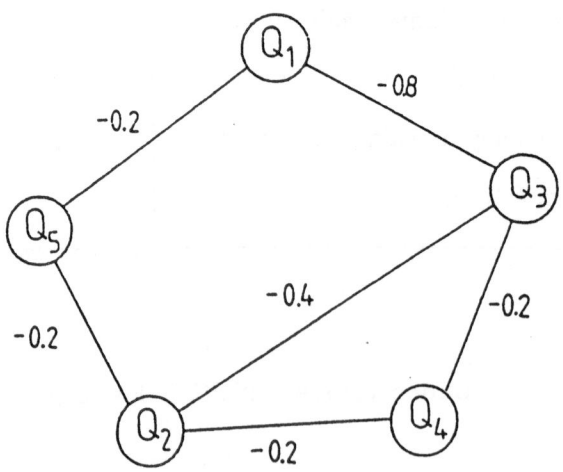

9.9 Selected Index Group Structure (NPTG Example)

of δ. The large negative median linkage between these two groups ($g_{13} = -0.7693$) indicates a conflict between the performance of the two channels. Also, as the two indices which indicate control effort (indices 2 and 3) are in groups which conflict with group 3, any improvement in the performance of δ may require a general increase in the control effort, apart from the sacrifice of the other channel's performance.

Design specifications were then drafted as a set of bounds for the indices as

$$c^{(0)} = (- 0.2, 200, 2000, 0.5, 1.0, 0.5, 150,$$
$$400, 1.0, 5.0, -5.0, -5.0)^T \tag{9.1.2.5}.$$

The attainability of the specifications was then analyzed using the interactive algorithm in section 7.3.

On first iteration, candidates for the group active members were identified as

$$Q_1^{(1)} = \{3, 11\}$$
$$Q_2^{(1)} = \{2\}$$
$$Q_3^{(1)} = \{13\} \tag{9.1.2.6}.$$
$$Q_4^{(1)} = \{1\}$$
$$Q_5^{(1)} = \{4\}$$

The scatter plot for the indices 3 and 11 was examined (fig. 9.10) and it was decided to select index 3 as the active member for group 1. To display the trade-off ([A2P2]-type) graph for the active indices, the permutation was obtained from the median linkage matrix using the heuristic algorithm in the appendix which yielded a group permutation as $(5, 1, 3, 2, 4)$. The implied permutation for the active members became $\mathcal{P}(\tilde{Q}^{(1)}) = (4, 3, 13, 2, 1)$. The graph was subsequently plotted as in

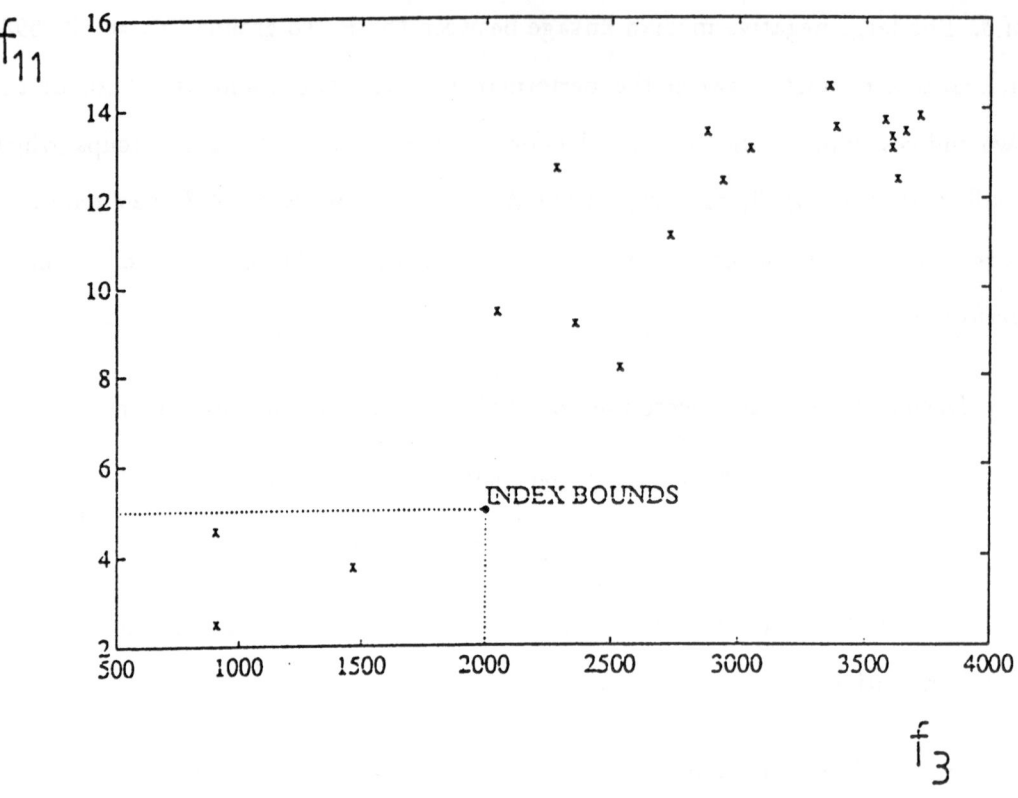

9.10 [A2P1]-Type Graph for Indices 3 and 11 in $Q_1^{(1)}$ (NPTG Example)

fig. 9.11a. It was decided that the bound for index 2 was to be relaxed from 200 to 450.

On second iteration, the active members were selected as $(11, 2, 13, 1, 4)$. The same group permutation was used which implied $\mathcal{P}(\tilde{Q}^{(2)}) = (4, 11, 13, 2, 1)$. From fig. 9.11b, it was decided to relax the bound for index 13 from -5 to -2 to avoid potential conflict with indices 11 and 2.

A further iteration of the algorithm gave $\mathcal{P}(\tilde{Q}^{(3)}) = (4, 11, 10, 2, 1)$ which revealed a potential conflict between indices 11 and 10 (fig. 9.11c). Index 11 was then relaxed from 5 to 7. No more critical conflicts were revealed in the subsequent iteration. The final set of bounds were therefore

$$c = (- 0.2, 450, 2000, 0.5, 1.0, 0.5, 150,$$
$$400, 1.0, 7.0, -5.0, -2.0)^T \tag{9.1.2.7}.$$

(3) Search Stage

An initial point was required for the numerical search to be conducted. One of the 19 candidates was selected for its satisfying the largest number of the set bounds while attaining relatively low values for the indices with the violated bounds. The search tools in section 6.2 were used, viz. moving boundary process as the dynamic minimax formulation (6.2.1.1) solved by simplex polytope method, monitored with the graphical display [P5]. A solution was found after stepping through 9 more candidate designs. Fig. 9.12 is the graphical display monitor of type [P5] (section 6.22) at termination of the stage.

(4) Evaluation Stage and Subsequent Design Decisions

The performance index values of the 9 new candidates found conformed rea-

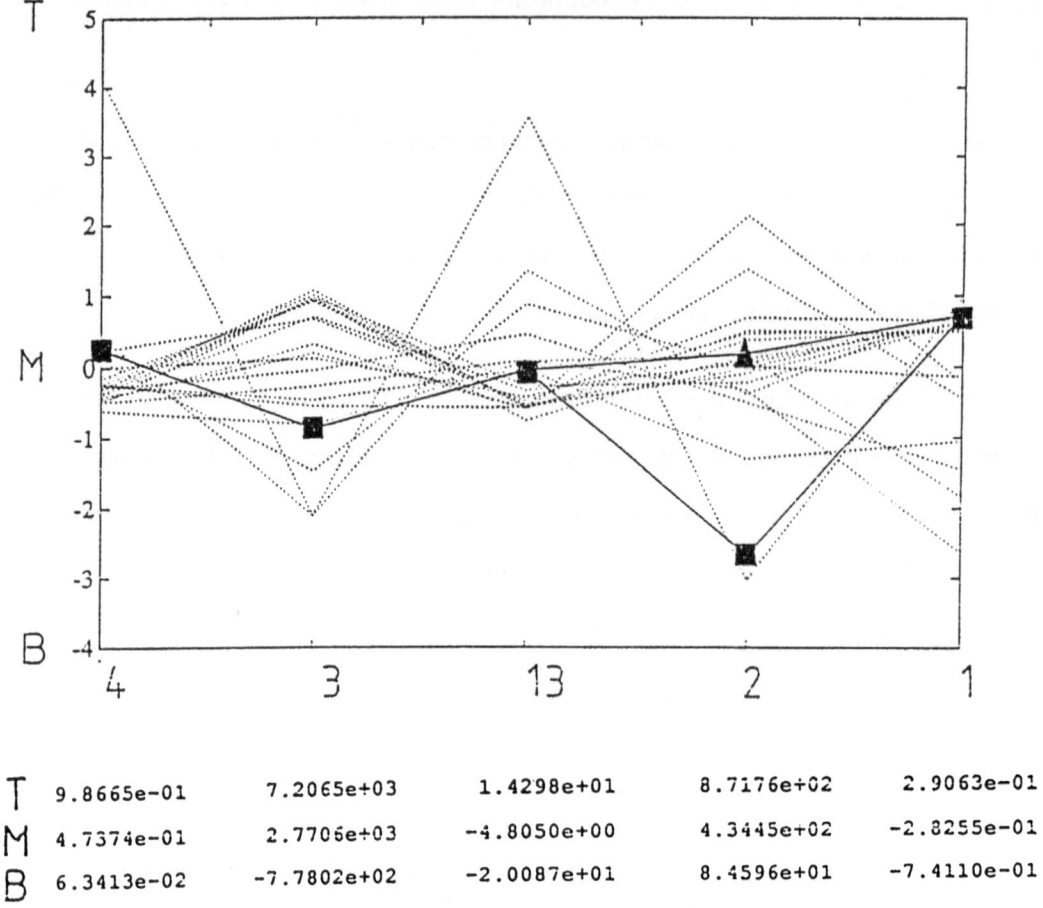

T	9.8665e-01	7.2065e+03	1.4298e+01	8.7176e+02	2.9063e-01
M	4.7374e-01	2.7705e+03	-4.8050e+00	4.3445e+02	-2.8255e-01
B	6.3413e-02	-7.7802e+02	-2.0087e+01	8.4596e+01	-7.4110e-01

9.11a Trade-Off Graph with $\mathcal{P}(\bar{Q}^{(1)})$ (NPTG Example)

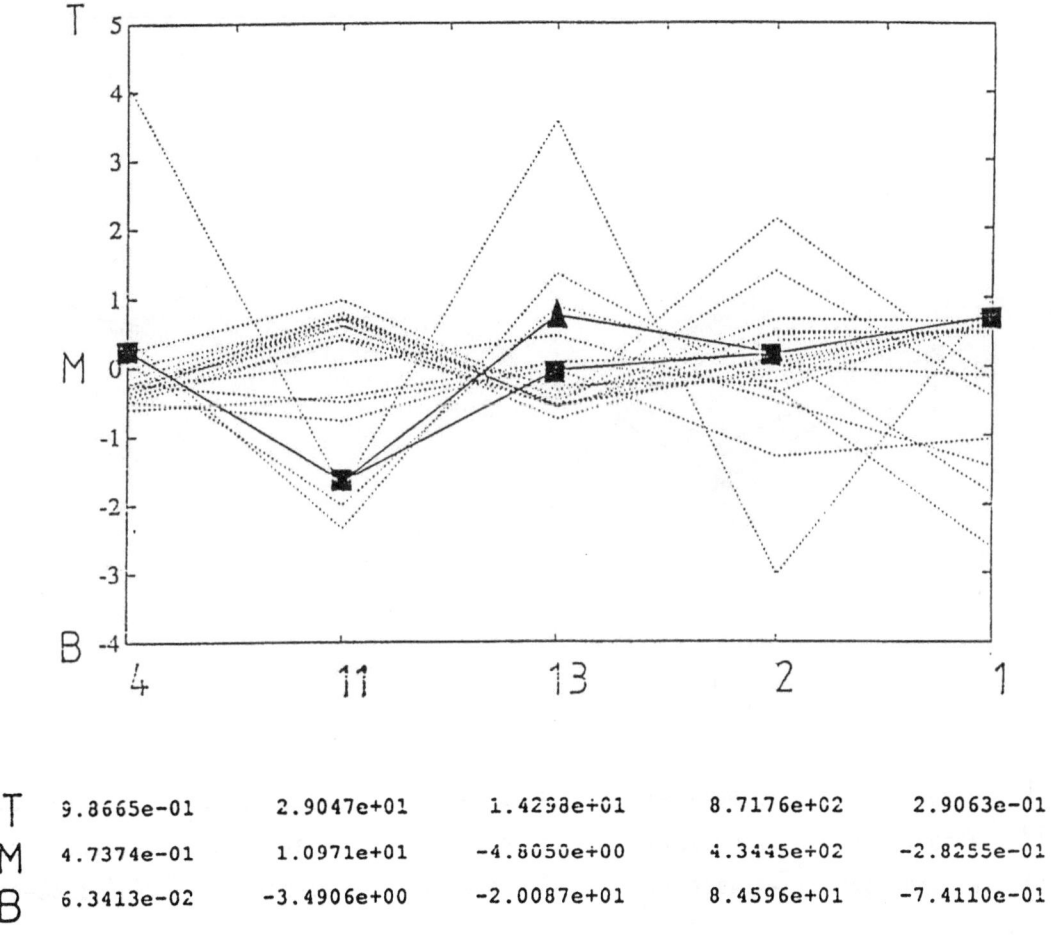

T	9.8665e-01	2.9047e+01	1.4298e+01	8.7176e+02	2.9063e-01
M	4.7374e-01	1.0971e+01	-4.8050e+00	4.3445e+02	-2.8255e-01
B	6.3413e-02	-3.4906e+00	-2.0087e+01	8.4596e+01	-7.4110e-01

9.11b Trade-Off Graph with $\mathcal{P}(\tilde{Q}^{(2)})$ **(NPTG Example)**

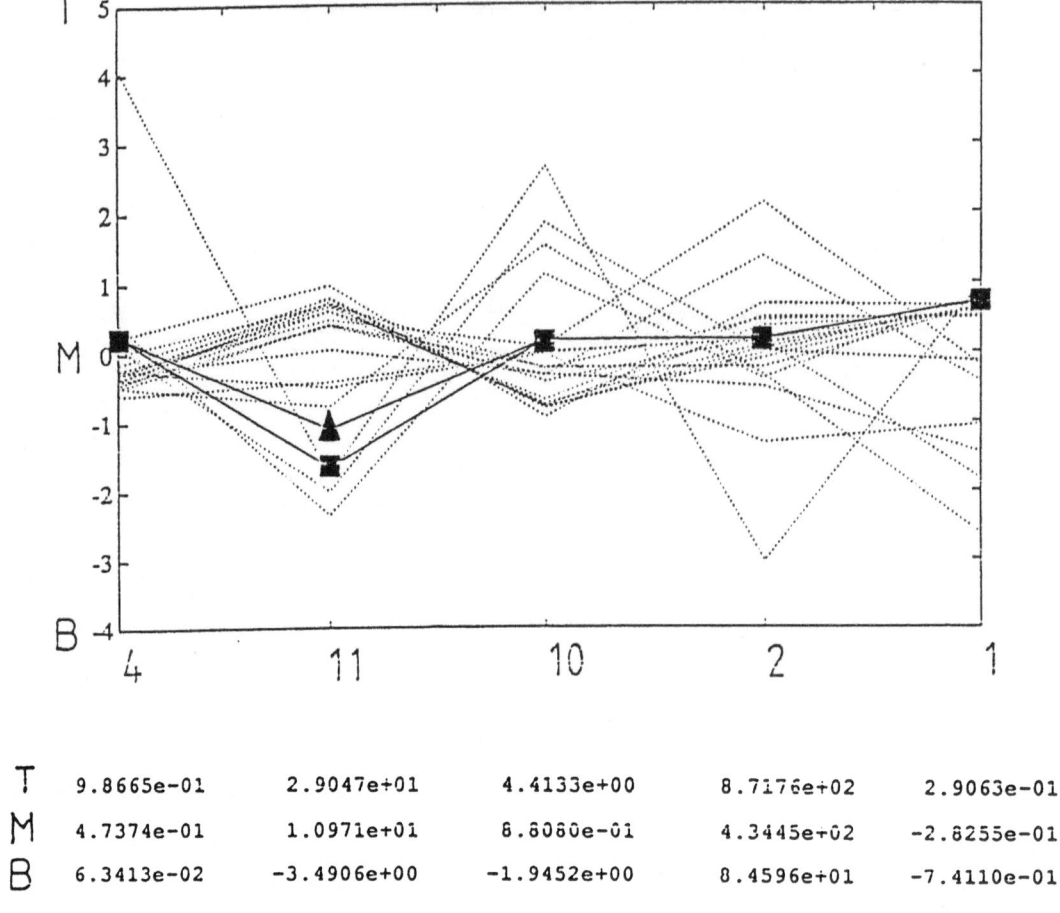

T	9.8665e-01	2.9047e+01	4.4133e+00	8.7176e+02	2.9063e-01
M	4.7374e-01	1.0971e+01	8.8080e-01	4.3445e+02	-2.8255e-01
B	6.3413e-02	-3.4906e+00	-1.9452e+00	8.4596e+01	-7.4110e-01

9.11c Trade-Off Graph with $\mathcal{P}(\tilde{Q}^{(3)})$ (NPTG Example)

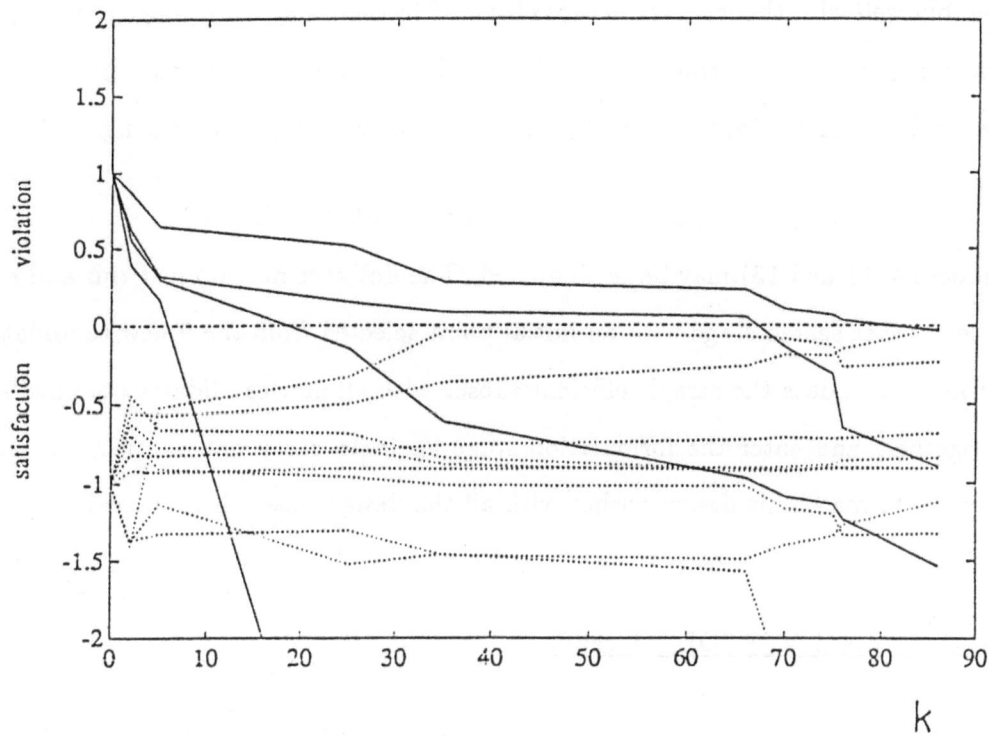

9.12 [P5]-Type Graphical Display Monitor (NPTG Example)

sonably well with the index structure obtained in the formulation stage (fig. 9.9). A trade-off graph with permutation $\mathcal{P}(\tilde{Q}^{(2)})$ (which contains all indices which bounds were relaxed in the formulation stage) was displayed in fig. 9.13 (c.f. fig. 9.11b).

There is evidence in both fig. 9.12 and fig. 9.13 that the three relaxed bounds (indices 4, 11 and 13) may be re-tightened. The designer may do just this and execute another search stage with an initial point selected from the 9 new candidates. Or he may update the sample efficient subset with all new candidates obtained (21 altogether) and enter the formulation stage again to carry out another full scale exercise to match his design wishes with all the design possibilities revealed.

9.2 A Typical V/STOL Aircraft

This problem is concerned with designing an LTI multivariable controller for the longitudinal motion of a typical V/STOL (Vertical/Short Take-Off and Landing) aircraft. The plant is highly non-linear due to very significant changes of the aircraft's dynamical behaviour over the whole flight envelope (the template of intended operating conditions on an altitude-speed graph). Therefore, the designed controller has to be robust over a useful range of flight conditions.

With LTI controllers, gain scheduling is a popular control strategy in which the controller coefficients are scheduled in a pre-programmed fashion according to the flight condition. Typically, the plant is linearized about a set of operating points. A set of LTI controllers with a common structure are then designed, one for each of the linearized models. Adaptive controllers are a promising alternative which may be considered a generalization of this scheme from the pre-programming of a finite number of scheduled coefficient sets to the continuous self-tuning of the coefficients.

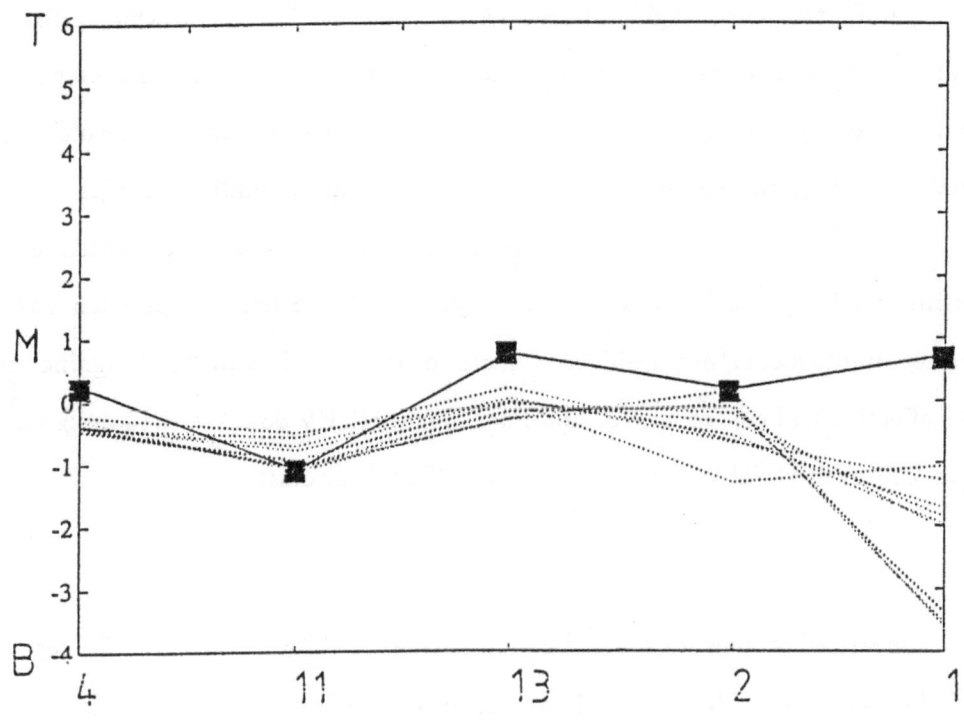

T	9.8665e-01	2.9047e+01	1.4298e+01	8.7176e+02	2.9063e-01
M	4.7374e-01	1.0971e+01	-4.8050e+00	4.3445e+02	-2.8255e-01
B	6.3413e-02	-3.4906e±00	-2.0087e+01	8.4596e+01	-7.4110e-01

9.13 Trade-Off Graph with $\mathcal{P}(\tilde{Q}^{(2)})$ for New Candidates Generated in the Search Stage (NPTG Example)

Yet another encouraging alternative appeared in [Kre 83] in which a fixed gain LTI controller was designed for a fighter aircraft over five operating points in the flight envelope. The simplicity of this scheme is unrivaled and the resulting flying qualities are reported as acceptable. Handling of a large number of design objectives was cited as the main difficulty. In actual fact, this was achieved with the vector optimization approach we looked at in chapter 3. Such an approach has been termed multi-model/multi-objective method as a viable tool to designing simple but effective LTI controllers for non-linear plants [DFV 85]. We shall design a fixed gain LTI multivariable controller for the V/STOL aircraft.

Linearized models for seven flight conditions ($FCOND = 1, 2, \ldots, 7$) were available (appendix C). The principal parameter for the flight conditions is longitudinal body velocity. For instance, $FCOND = 1$ is the hovering condition when the longitudinal body velocity is 8.4 ft/s while $FCOND = 7$ is when the aircraft is at the cruising speed of 506 ft/s. For $FCOND = 1, 2, 3, 4$ which correspond to low speeds, there are three inputs (U1, U2, U3) and ten states (X1, X2, ..., X10) (fig. 9.14). States X1 to X4 are states of aerodynamics, X5 - X7 are actuator dynamics while states X8 - X10 are engine states. For $FCOND = 5, 6, 7$, the nozzle is at full aft position and U3 ceases to be a usable input for the linearized model. Similarly, its corresponding actuator dynamics state X7 is dropped from the set of states. A common structure exists across all models in which the engine states are principally affected by U2 while the aerodynamic states were affected by U1 and U3 (appendix C). The coupling from the engine states to the aerodynamic states is significant while the that in the reverse direction is negligible.

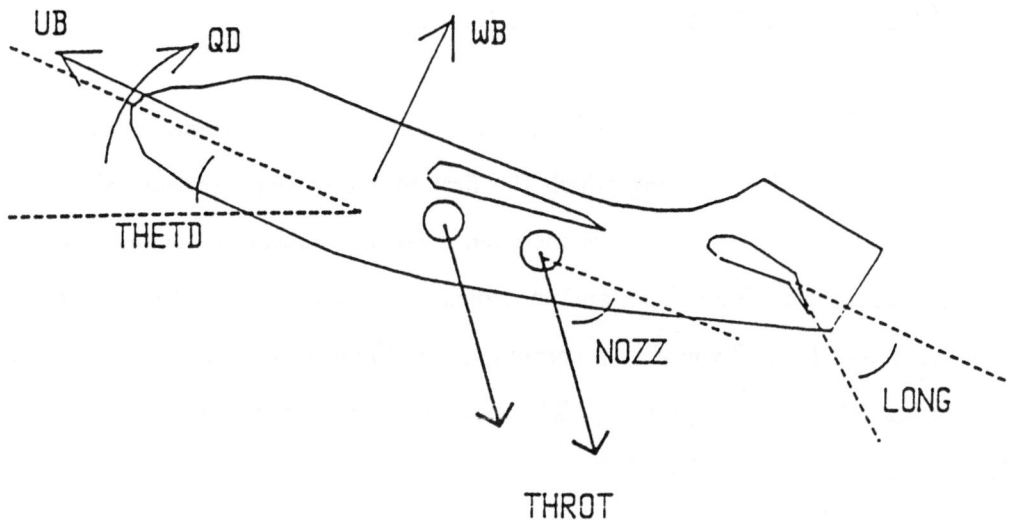

STATES

THETD	(X1)	pitch attitude in degrees
QD	(X2)	pitch rate in deg/s
UB	(X3)	longitudinal body velocity in ft/s
WB	(X4)	normal body velocity in ft/s
LONG	(X5)	tailplane angle in degrees
THROT	(X6)	normalized throttle setting (0.0 - 1.0)
NOZZ	(X7)	nozzle angle in degrees
FNP	(X8)	fan speed
HNP	(X9)	compressor speed
QEF	(X10)	fuel flow

INPUTS

ALONG	(U1)	tailplane angle command in deg/s
ATHROT	(U2)	normalzed throttle setting command (0.0 - 1.0)
ANOZZ	(U3)	nozzle angle command in degrees

9.14 A Typical V/STOL Aircraft

9.2.1 Level I: Innovation - Feasible Set Generation

Decoupled control was desired for pitch hold and longitudinal velocity hold for all flight conditions, among which the first four also required normal velocity hold. Therefore, design objectives included close loop stability, dynamic performance (tracking and decoupling) and control effort, while the states to be controlled were X1, X3 and X4. Typical step commands for them were given as 5.0, 25.0 and 10.0 for X1, X3 and X4 respectively. Therefore, we normalize them as $Y1 = X1/5.0$, $Y2 = X3/25.0$ and $Y3 = X4/10.0$.

A design decision to be made was for what range of flight conditions a fix gain controller was to be designed. It was decided that only the three-input low speed models were to be considered.

Asymptotic tracking was considered desirable. Therefore, a multivariable (3-input-3-output) proportional and integral controller for output feedback of Y1, Y2 and Y3 was to be used. Pitch rate (X2) is a popular state for feedback for longitudinal motion control due to the achievable high quality of measurement. Therefore, a constant gain partial state feedback of the X2 together with the three outputs (X1, X3, X4) was also used. The resulting controller configuration is illustrated in fig. 9.15. All coefficients in the controller gain matrices were used as design parameters, of which there were 30.

No analytical methods were available in constructing an initial design with the suggested structure. Instead, a few optimizations were formulated in an *ad hoc* manner to obtain the individual controller components in fig. 9.15, which were combined to give an initial design.

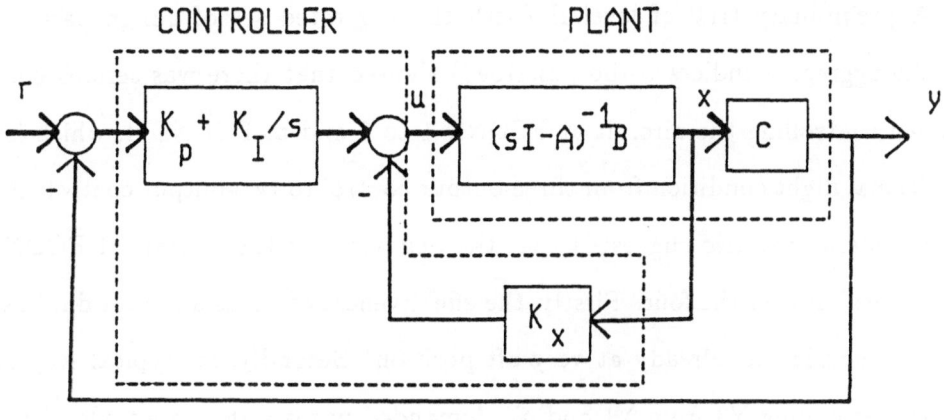

$$u = (K_p + K_I/s)(r-y) - K_x \begin{pmatrix} x_1 \\ x_2 \\ x_3 \\ x_4 \end{pmatrix}$$

$$K_p, \ K_I \qquad R^{3x3}$$
$$K_x \qquad R^{3x4}$$

9.15 Controller Configuration (V/STOL Problem)

A preliminary trial of level II (with the suggested set of design parameters and the aggregate indices to be described) showed that there was serious conflict between controlling the aircraft at FCOND 1 to 3 and at FCOND 4, which is the transitional flight condition from three-output control to two-output control. Physical considerations also suggested that the behaviour of the aircraft at FCOND 4 was atypical among the four. Firstly, the effectiveness of U3 as an input diminished since the nozzle was already at very aft position. Secondly, for typical step commands, decoupling Y3 from Y1 and Y2 demanded unrealizable input effort (which was be appreciated by comparing the inverse of the steady state transfer function matrix with input saturation levels). FCOND 4 was subsequently dropped from the set of flight conditions for which a fixed gain controller was to be designed.

9.2.2 Level II: Trade-offs - Matching Wishes and Possibilities

(1) Exploratory Stage

Five aggregate performance indices were formulated for each of the three flight conditions as

$$\hat{f}_i : \Re^{30} \rightarrow \Re, \quad \hat{i} \in \hat{Q} = \{1, 2 \ldots, 15\}, \tag{9.2.2.1}$$

the details of which are described in fig. 9.16.

6 design iterations were executed with different sets of good and bad values. To simplify the handling of all the thirty good and bad values, those for the indices which indicated the same design objective for different flight conditions were varied together. In other words, the relationships among objectives *over* all three flight conditions were explored, instead of those *between* them. Fig. 9.17 gives a brief account of the design iterations executed. As a result, 118 distinct controllers were produced which values of the aggregate indices were plotted in fig. 9.18.

i #	INDICATES
1, 6, 11	control loop stability
2, 7, 12	damping
3, 8, 13	control effort
4, 9, 14	dynamic coupling
5, 10, 15	transient speed

1,2,3,4,5 ———— FCOND 1.

6,7,8,9,10 ———— FCOND 2.

11,12.13,14,15 —— FCOND 3.

9.16 Aggregate Performance Indices (V/STOL Problem)

I	INTEND FURTHER TO	AT FURTHER EXPENSE OF
1	reduce coupling	more control effort
2	reduce control effort	all other indices
3	reduce control effort	more coupling
4	reduce control effort and coupling	slower transient speed
5	reduce control effort and coupling	less stability
6	reduce control effort and coupling	all other indices

I - Design Iterations

9.17 Design Iterations in the Exploratory Stage (V/STOL Problem)

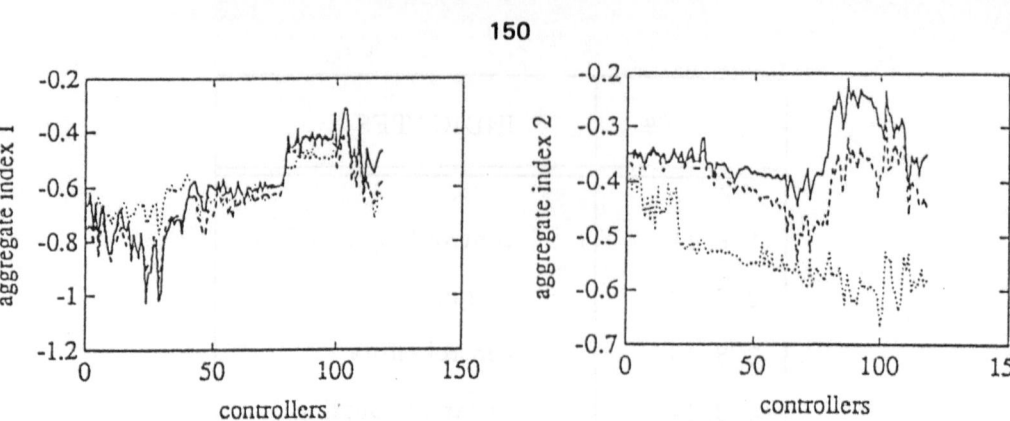

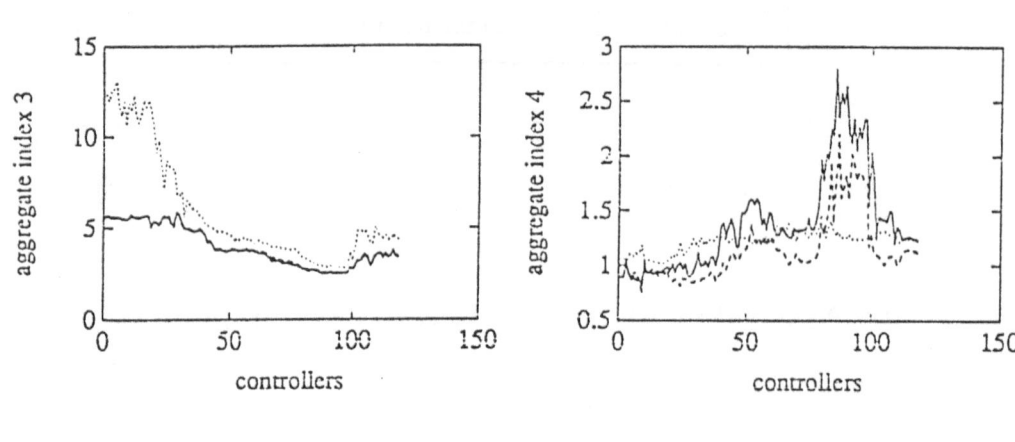

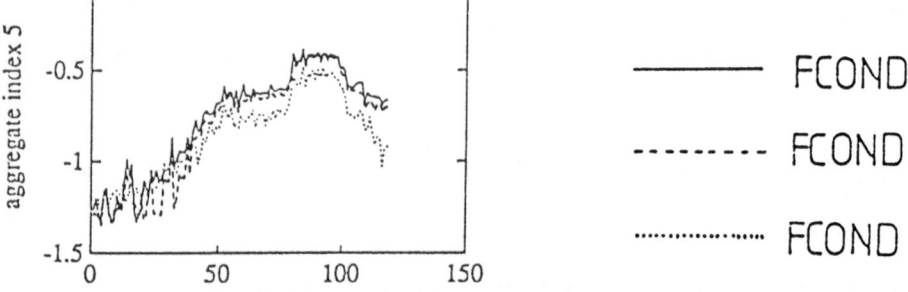

9.18 Aggregate Performance Indices' Values (V/STOL Problem)

An elaborated set of performance indices were chosen, of which there were 69 altogether :

$$f_i : \Re^{30} \to \Re, \quad \hat{i} \in Q = \{1, 2 \ldots, 69\}, \tag{9.2.2.2}$$

with 23 for each flight condition. They are described in fig. 9.19. All 118 candidates were not disproved of their efficiency.

(2) Formulation Stage

Choosing $s_r = 0.3$ gave 7 groups as :

$$Q_1 = \{3, 4, 5, 6, 14, 15, 16, 26, 27, 28, 29, 33,$$

$$37, 38, 39, 48, 49, 50, 51, 52, 56, 60, 62, 67\}$$

$$Q_2 = \{7, 8, 17, 30, 31, 40, 53, 54, 63\}$$

$$Q_3 = \{2, 12, 18, 20, 21, 22, 25, 41, 43, 44, 45, 66\}$$

$$Q_4 = \{1, 9, 13, 24, 32, 35, 36, 47\} \tag{9.2.2.3}$$

$$Q_5 = \{23, 46, 55, 65, 69\}$$

$$Q_6 = \{10, 11, 19, 34, 42, 59\}$$

$$Q_7 = \{57, 58, 61, 64, 68\}$$

The median linkage matrix is

$$G = \{g_{l_1 l_2}\} =$$

$$\begin{pmatrix}
0.9018 & 0.5130 & -0.0877 & -0.7349 & -0.3806 & -0.1453 & 0.6362 \\
0.5130 & 0.8037 & -0.0929 & -0.4361 & 0.1182 & 0.1482 & 0.3617 \\
-0.0877 & -0.0929 & 0.6423 & 0.2082 & 0.4381 & -0.4580 & 0.3707 \\
-0.7349 & -0.4361 & 0.2082 & 0.7715 & 0.3850 & 0.1060 & -0.3782 \\
-0.3806 & 0.1182 & 0.4381 & 0.3850 & 0.7841 & -0.1767 & -0.0865 \\
-0.1453 & 0.1482 & -0.4580 & 0.1060 & -0.1767 & 0.7971 & -0.4146 \\
0.6362 & 0.3617 & 0.3707 & -0.3782 & -0.0865 & -0.4146 & 0.7876
\end{pmatrix}$$

$$\tag{9.2.2.4}$$

Assigning $s_c = -0.3$ gave 6 conflicting pairs. The resulting index structure is displayed in fig. 9.20. It may be expected that a physical interpretation of the

i #	DEFINITION	INDICATES
1, 24, 47	maximum real part of the close loop poles	control loop stability
2, 25, 48	negative of minimal damping coefficient close loop poles	damping
3, 26, 49	maximum overshoot (step in Y1) of Y1	damping
4, 27, 50	IAE of step (in Y1) response of Y1	command tracking
5, 28, 51	maximum overshoot (step in Y2) of Y2	damping
6, 29, 52	IAE of step (in Y2) response of Y2	command tracking
7, 30, 53	maximum overshoot (step in Y3) of Y3	damping
8, 31, 54	IAE of step (in Y3) response of Y3	command tracking
9, 32, 55	maximum absolute value of step (in Y1) response of U1	control effort
10, 33, 56	maximum absolute value of step (in Y1) response of U2	control effort
11, 34, 57	maximum absolute value of step (in Y1) response of U3	control effort
12, 35, 58	maximum absolute value of step (in Y2) response of U1	control effort
13, 36, 59	maximum absolute value of step (in Y2) response of U2	control effort
14, 37, 60	maximum absolute value of step (in Y2) response of U3	control effort
15, 38, 61	maximum absolute value of step (in Y3) response of U1	control effort
16, 39, 62	maximum absolute value of step (in Y3) response of U2	control effort
17, 40, 63	maximum absolute value of step (in Y3) response of U3	control effort
18, 41, 64	maximum absolute value of step (in Y1) response of Y2	coupling
19, 42, 65	maximum absolute value of step (in Y1) response of Y3	coupling
20, 43, 66	maximum absolute value of step (in Y2) response of Y1	coupling
21, 44, 67	maximum absolute value of step (in Y2) response of Y3	coupling
22, 45, 69	maximum absolute value of step (in Y3) response of Y1	coupling
23, 46, 69	maximum absolute value of step (in Y3) response of Y2	coupling

#	1, 2, ..., 23	———————	FCOND 1.
	24, 25, ..., 46	———————	FCOND 2.
	47, 48, ..., 69	———————	FCOND 3.

9.19 Performance Indices (V/STOL Problem)

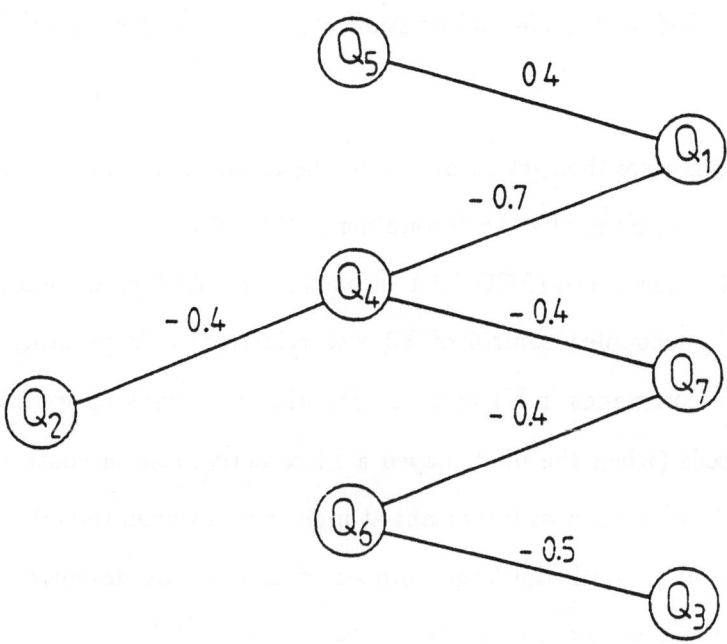

9.20 Selected Index Group Structure (V/STOL Problem)

stucture would yield very useful information toward the understanding the design problem.

Index bounds were then set according to the following designer's wishes. Coupling considerations, especially the decoupling of Y3 from commands in Y1 and Y2, were important at low speed (FCOND 1 in particular). At higher speeds (FCOND 3 in particular), decoupled control of Y3 was relatively unimportant, so was its decoupling from commands in Y1 and Y2. Stability and tracking were less important at low speeds (when the pilot played a more active role in control). Control effort considerations were most important at high speeds (when the pilot gave more step-like commands). With the large number of indices, the designer was able to express all these wishes in the index bound values (fig. 9.21).

The attainability of the bounds was then analyzed using the interactive algorithm in section 7.3. The first iteration identified the candidates for the group active members as

$$Q_1^{(1)} = \{16, 37, 39, 56, 60, 62\}$$

$$Q_2^{(1)} = \{54\}$$

$$Q_3^{(1)} = \{21\}$$

$$Q_4^{(1)} = \{13\} \qquad (9.2.2.5).$$

$$Q_5^{(1)} = \{65\}$$

$$Q_6^{(1)} = \{19, 59\}$$

$$Q_7^{(1)} = \{64\}$$

After active indices were selected, a liberal choice of alternative permutations were used to display a set of trade-off ([A2P2]-type) graphs of them. A careful search for index pairs with critical conflicting bounds were conducted. Bounds were relaxed accordingly. Four conflicting pairs were identified in the first iteration ([54,19],

FCOND 1	FCOND 2	FCOND 3	DESIGN OBJECTIVES
(1) -0.3000	(24) -0.5000	(47) -0.5000	control loop stability
(2) -0.3000	(25) -0.3000	(48) -0.3000	damping
(3) 0.2000	(26) 0.2000	(49) 0.2000	damping
(4) 5.0000	(27) 4.0000	(50) 3.5000	command tracking
(5) 0.2000	(28) 0.2000	(51) 0.2000	damping
(6) 5.0000	(29) 4.0000	(52) 3.5000	command tracking
(7) 0.2000	(30) 0.2000	(53) 0.2000	damping
(8) 5.0000	(31) 4.0000	(54) 3.5000	command tracking
(9) 15.0000	(32) 10.0000	(55) 5.0000	control effort
(10) 0.2000	(33) 0.3000	(56) 0.3000	control effort
(11) 30.0000	(34) 20.0000	(57) 20.0000	control effort
(12) 15.0000	(35) 10.0000	(58) 5.0000	control effort
(13) 0.2000	(36) 0.3000	(59) 0.3000	control effort
(14) 30.0000	(37) 20.0000	(60) 20.0000	control effort
(15) 15.0000	(38) 10.0000	(61) 5.0000	control effort
(16) 0.2000	(39) 0.3000	(62) 0.3000	control effort
(17) 30.0000	(40) 20.0000	(63) 20.0000	control effort
(18) 0.1000	(41) 0.1000	(64) 0.1000	coupling
(19) 0.1000	(42) 0.2000	(65) 0.5000	coupling
(20) 0.2000	(43) 0.2000	(66) 0.2000	coupling
(21) 0.1000	(44) 0.2000	(67) 0.5000	coupling
(22) 0.2000	(45) 0.2000	(68) 0.5000	coupling
(23) 0.1000	(46) 0.1000	(69) 0.5000	coupling

9.21 Index Bounds Specified in the Formulation Stage (V/STOL Problem)

[19,21], [16,13] and [13,54]) which led to the relaxation of bounds for indices 54 (3.5 to 4.0), 19 (0.1 to 0.2) , 13 (0.2 to 0.3) and 21 (0.1 to 0.2). In the second iteration, five pairs were identified ([62,59], [59,31], [21,59], [65,62] and [62,36]) which led to the relaxation of bounds for indices 59 (0.3 to 0.4), 65 (0.5 to 0.6), 36 (0.3 to 0.35) and 31 (0.4 to 0.45).

(3) Search Stage

When the dynamic minimax formulation (6.2.1.1) was used, it was not difficult to appreciate the complexity of the resulting auxiliary optimizations. The function being optimized was non-smooth and non-linear, and there were 30 design parameters. We therefore decided to help in whatever ways we could in the computer's numerical search using the simplex polytope method, and the followings were done :

(1) Instead of selecting just one candidate design as the initial point, 31 were selected to compose an initial simplex polytope in the 30-dimensional design parameter space. A selection criterion used was the number of satisfied bounds for the candidate group active members in (9.2.2.5). The best candidate violated 13 of the 69 index bounds.

(2) Alternative formulations to generate candidate solutions (such as (6.2.2)) for the search stage were executed in parallel as secondary search processes. Any promising candidates generated were included in the simplex polytope of the dynamic minimax formulation. In this way, the numerical search process of the dynamic minimax formulation acted as the principal search mechanism whose attention was constantly drawn to "good regions" in the design parameter space identified by the secondary searches.

The designer was well capable of providing such helps primarily because of the simplicity of the simplex polytope method as well as the availability of the evaluation tools, with which he looked for promising candidates in trajectories traversed in the secondary search processes. Such mode of co-operation was found to be an effective means to combine the strengths of both the human designer and the computer in the numerical search.

(4) Evaluation Stage and Subsequent Design Decisions

The search was terminated when 2 bounds remained violated. They were index 56 (= 0.482) and index 62 (= 0.305). Both were control effort consideration of the throttle setting (U2) at FCOND 3. The step time responses are plotted in fig. 9.22. The gain matrices of the final controller are :

$$K_P = \begin{pmatrix} -3.6975 & 0.5195 & 3.6127 \\ 0.3255 & 0.1440 & -0.1622 \\ -2.4063 & -8.1211 & 1.8965 \end{pmatrix}$$

$$K_I = \begin{pmatrix} -11.3619 & 1.0541 & -0.4024 \\ 0.7292 & 0.1101 & -0.3026 \\ -5.2248 & -24.9775 & -0.1019 \end{pmatrix} \quad (9.2.2.6)$$

$$K_X = \begin{pmatrix} -4.3857 & -2.1774 & -0.0693 & -0.3027 \\ 0.3410 & 0.1066 & 0.0267 & -0.0818 \\ -0.1267 & -4.8179 & -1.9217 & -1.2358 \end{pmatrix}$$

Fig. 9.23 is the graphical display monitor of type [P5] (section 6.22) at termination of the search stage. Although there are 69 graphs plotted together, the pattern is clear enough for the designer to have reasonable understanding of the progress. He can appreciate the effort spent in the satisfaction of the bounds as well as predict the effort required in further progress. As shown in fig. 9.23, the designer might have decided to have a closer look at the new candidates generated. He might have gone to the customer and come back with modified wishes, re-iterating some of the previous stages to conduct another search.

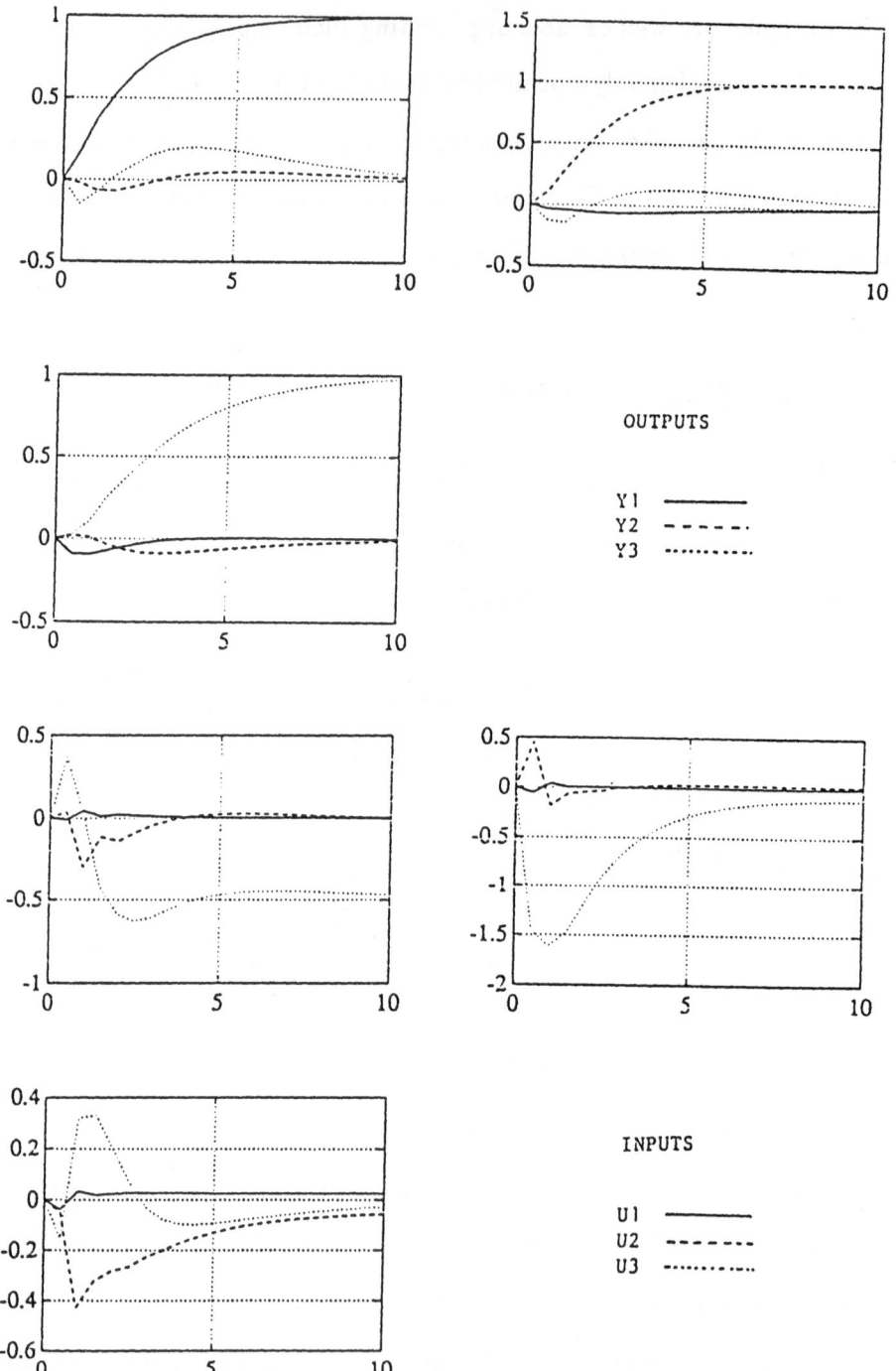

9.22a Time Responses of Final Design (V/STOL Problem: FCOND 1)

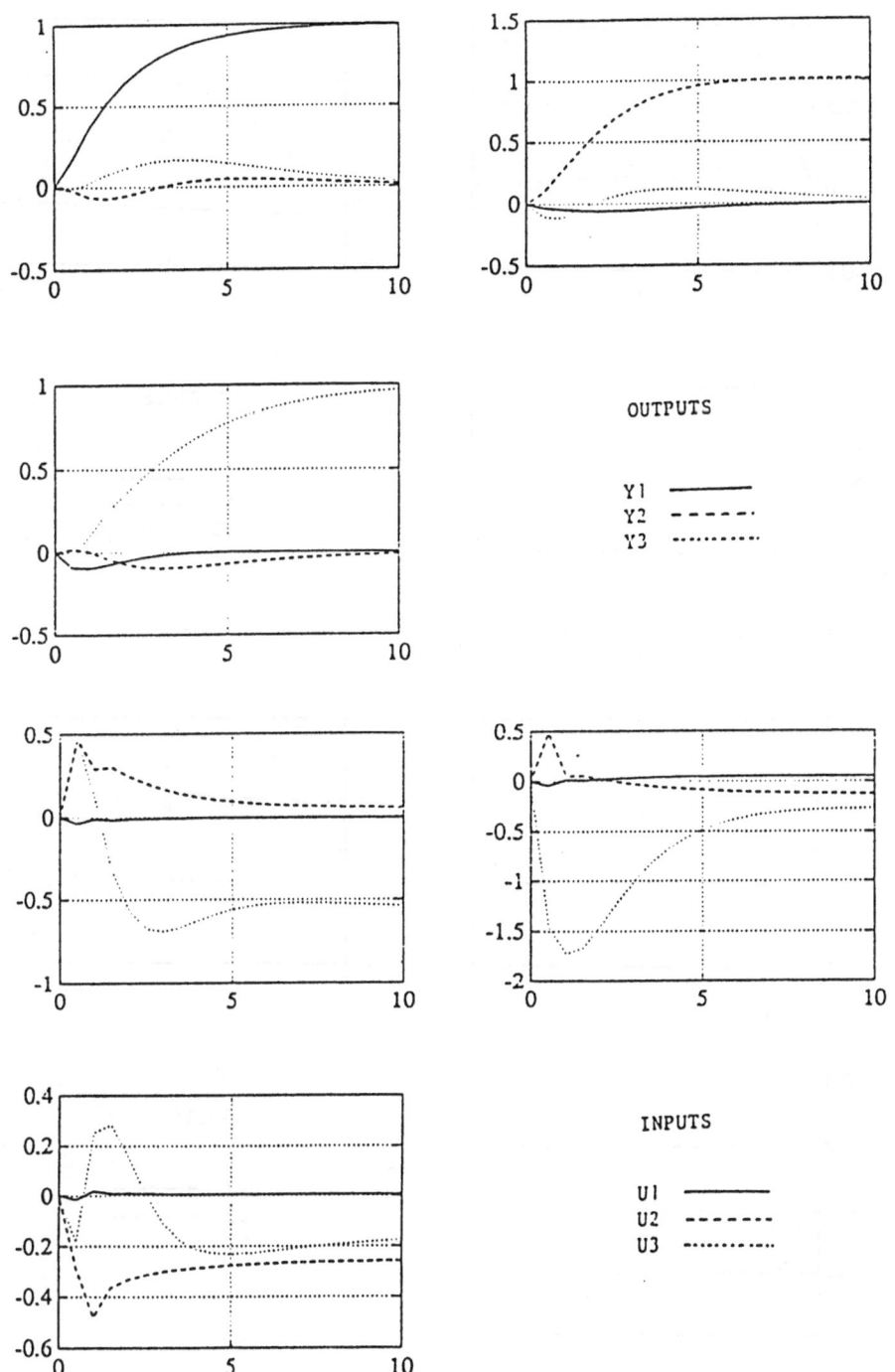

9.22b Time Responses of Final Design (V/STOL Problem: FCOND 2)

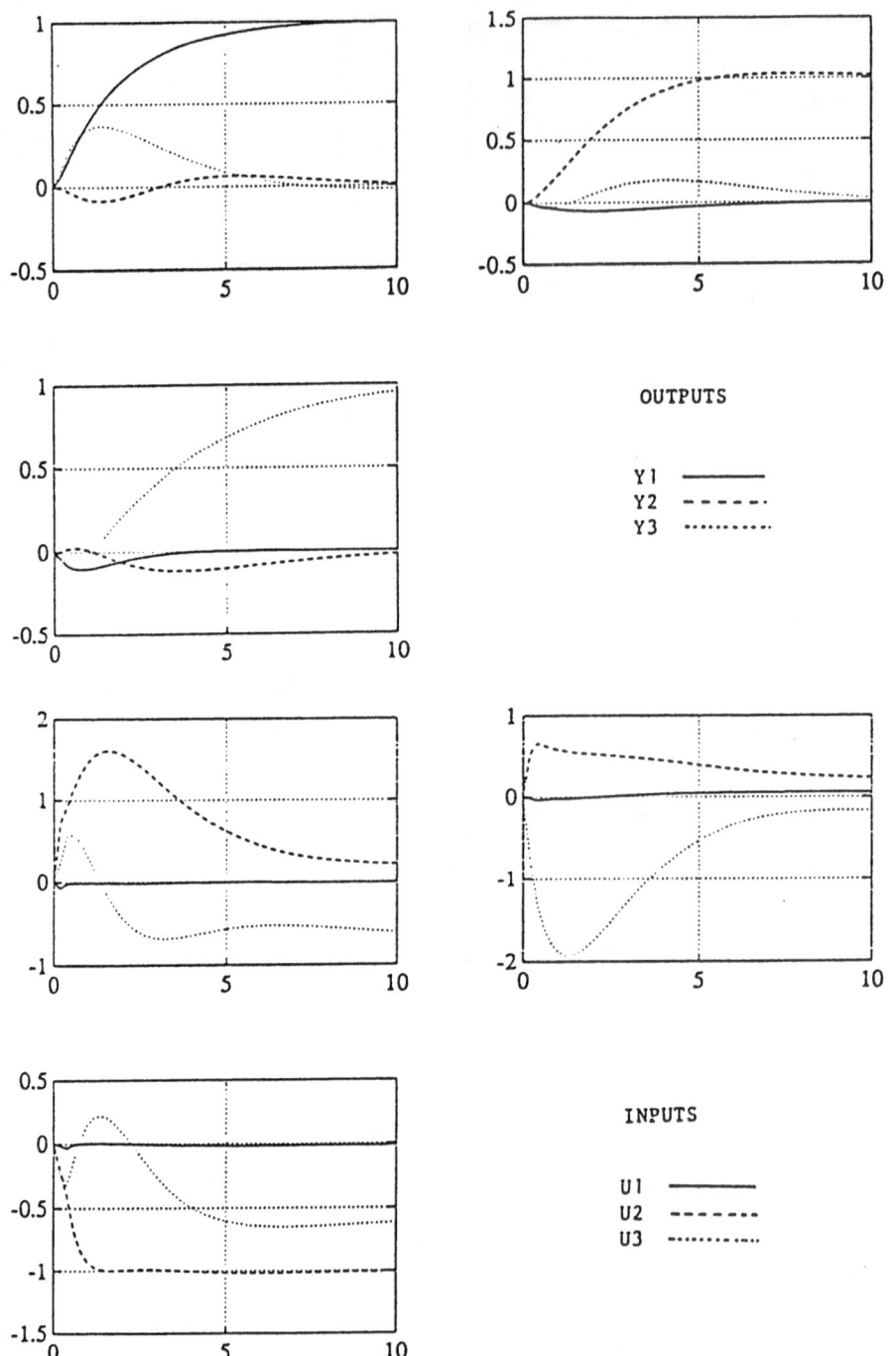

9.22c Time Responses of Final Design (V/STOL Problem: FCOND 3)

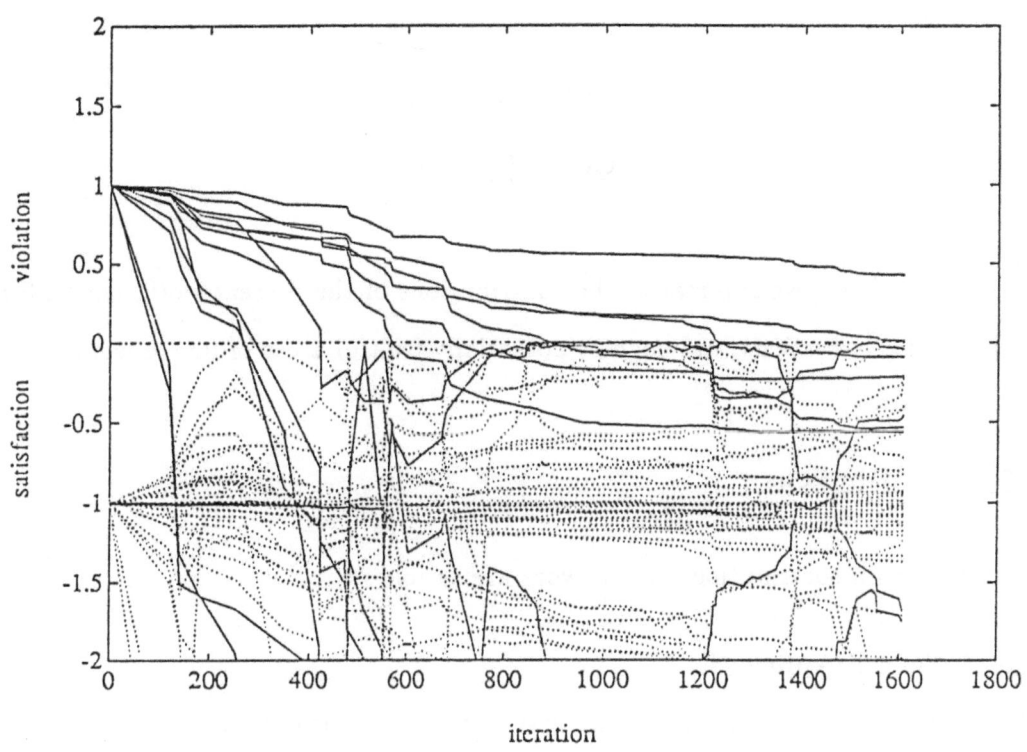

9.23 [P5]-Type Graphical Display Monitor (V/STOL Problem)

CHAPTER 10

CONCLUSION

To conclude, we summarize the contributions of the present work, note a few important observations, and suggest potentially fruitful areas of future work.

10.1 Contributions

The main contributions of this work are as follows :

(1) A Search Approach

The work conducted is probably the first integrated approach to designing control systems by search, which enables the principal design stages (formulation, generation and evaluation) to be supported in a co-ordinated as well as most general manner. To computer-aid this design approach, interactive multi-objective programming has been shown to be a suitable framework for the computer. The framework provides the level of abstraction as well as the organization of the CACSD facilities. As a result, design tools have been identified, and prototypes implemented in the popular CACSD environment Pro-Matlab.

(2) A Design Strategy

A software tool approach to supporting man-machine interaction with interactive algorithms has been demonstrated to be a flexible means of communication between the designer and the computer. In our case, novel interactive algorithms have been developed which enable the designer to comprehend design possibilities as well as to match his wishes with them. The various simple but well-designed graphs have been found to play an important role. Since the designer's toolbox is enriched with such interactive tools, a design strategy has been developed which ensures an effective and co-ordinated use of the design tools.

(3) Control System Design

In the Pro-Matlab prototype, the major effort required from the designer is the writing of a computer program which takes the design parameters as input and returns the performance indices. Understanding of the design method (or the canonical form) which underlies the parametrization is preferable (it helps the construction of initial designs in level I), *but not crucial*. The search approach allows him to concentrate directly on the design objectives. The usability of potentially complicated design methods is enhanced. This is especially true in the case of multivariable control system design.

10.2 Observations

(1) Data Visualization

This work has demonstrated the power of human perception. Simple but well-designed two-dimensional graphical plots have proved to be a sufficient means for the human designer to perceive high-dimensional information such as the relations among multiple performance indices. However, colour graphics will help in resolution and differentiation.

(2) Computer-Aided Design

Different design methods may be liberally used (in level I) whose contributions may be effectively combined, using the numerical search approach in conjunction with the maintenance of the sample efficient subset (in level II).

(3) Decision Support System Approach

The decision support system approach to scientific management [Spr 82] argues that informed decision-makers are better decision-makers. The author's own experience has testified this in the case of CACSD: *informed designers are better designers.* The proper role of a computer may be summarized as the designer's information assistant (similar to the information manager in a business organization).

10.3 Suggestions for Future Works

It is important for the numerical search used in level II of the strategy to have uninhibited access to the constructed feasible sets. For this, we need :

(1) Effective Parametrizations

Boyd et. al suggested the use of finite dimensional subspaces of the infinite dimensional Q-parametrization of all stabilizing controllers in numerical optimizations [Boy 88]. The Q-parametrization is convex and so are the subspaces they suggested. Glover and Doyle [Glo 88] parametrized a class of suboptimal H^∞ controllers, which are stabilizing as well as satisfying an H^∞-norm bound. Finite dimensional subspaces need to be constructed if numerical searches are to be used.

(2) Effective Aggregate Indices

A set of effective aggregate indices are needed for a good approximation of the conventional control system design objectives in the exploratory stage of Level II. A good starting point may be the primary indicators due to Pang and MacFarlane [Pan 87]. Boyd et. al [Boy 88] noted that most, if not all, common close loop specifications are convex functions of the Q-parametrization. The resulting optimization formulations can therefore be solved effectively by numerical means, i.e. solutions are found whenever they exist.

(3) Effective Optimizations

The formulated search problems (the minimax and the inequalities formulations) are always non-smooth. [Kie 85] and [Fle 87] reported the subgradient methods as a promising approach to non-smooth optimization problems. Experimentation is needed in their application. However, it is also desirable to keep the polytope method for its good sampling property. Mixed mode operations of this and the subgradient methods may be the direct way of having the benefits of both.

To further empower the human designer as well as to enhance his communication with the computer, the followings are useful :

(4) Parallel Optimizations

Parallel optimizations are an effective way to conduct the secondary search processes as suggested in section 9.2.2. Tools for managing the resulting structured searches over the parameter space are required.

(5) Advanced Interface Facilities

This is a most general requirement for all computer-assisted activities. The particular concerns of the design approach and the strategy developed point to the need of a direct manipulation interface [Nor 86] and true interactive graphics capabilities (graphical output *and* input). Phaal has demonstrated a strong case for the object-oriented approach of the Smalltalk programming environment [Pha 87].

(6) Database Facilities

An integrated design database facility for conventional (tables) as well as unconventional engineering data (e.g. design history, linear systems) is invaluable to the designer [Mac 87]. It can extend the data manipulation facilities suggested in chapter 8 to all design stages, the co-ordination of which will be greatly enhanced. Of particular importance is the expressive power available in the data sublanguage. The DL language developed by Breuer [Bre 88] is a powerful functional language with all database manipulation capabilities. It is interesting to note that it has the full expressive power of the evaluation language suggested in chapter 8.

APPENDIX A

TRADE-OFF GRAPH PERMUTATIONS

We consider the problem of obtaining a good permutation of a set of group active indices, $\tilde{Q} = \{i_l, l \in M = \{1, 2, \ldots, m\}\} \in Q$ from a given linkage matrix $\Lambda = \{\lambda_{l_1 l_2}\} \in \Re^{m \times m}$ of the indices ($\lambda_{l_1 l_2}$ being a measure of linkage between group active indices i_{l_1} and i_{l_2}), which is either the group median linkage matrix G in (7.2.4), or the submatrix of the correlation matrix S in (7.2.1) which correspond to the i_l's. The more negative is $\lambda_{l_1 l_2}$, the larger is the conflict between i_{l_1} and i_{l_2}.

The primary purpose of the permutation is for constructing a trade-off graph which displays visual patterns to enhance the designer's understanding of the conflicts among the indices. Therefore, we consider the following as two criteria for a good permutation : (i) most of the neighbours should be conflicting and (ii) the most serious conflicting pairs should be neighbouring to each other. (If G is used, all trade-off graphs generated in the algorithm of section 7.2 will have the same permutation. Relationships among index groups may manifest as common patterns which exist in all the graphs. If S is used instead, the relationships among the active indices selected each time are examined more closely, when the permutation depends on their rank correlations.)

If we consider the linkages as distances between the active indices and seek to minimize the sum of all distances, the problem is equivalent to a travelling salesman's problem. Since the number of active indices is not large, we can afford solving this problem to obtain the resulting permutation, which may be expected to be a good one.

As a simple alternative, we devise the following heuristic procedure which constructs the permutation by progressively building up an ordered list of the group active members.

[**STEP 1**] Initialize an ordered list as $\pi^{(0)} = [i_{l_1}; i_{l_2}]$ where i_{l_1} and i_{l_2} are the closest index pair (i.e. $\lambda_{l_1 l_2}$ is minimum among entries in Λ. Let $k = 0$.

[**STEP 2**] Among the indices which are not in list $\pi^{(k)}$, select the one which is closest to either the first or last of the list. Add the index to the list to obtain $\pi^{(k+1)}$: prepend if closest to the first, append otherwise.

[**STEP 3**] Set $k = k + 1$. Terminate if $k = m$. Otherwise go to step 2.

We may expect this greedy algorithm to perform well when m is not too large, which is always our case. It is simple, and often put seriously conflicting pairs next to each other (step 1 ensures this for the most conflicting pair).

APPENDIX B

LINEAR MODEL OF A NUCLEAR-POWERED TURBO-GENERATOR

A

```
 0.00e+00  1.00e+00  0.00e+00  0.00e+00  0.00e+00  0.00e+00  0.00e+00  0.00e+00  0.00e+00  0.00e+00
 0.00e+00 -1.13e-01 -9.81e-01 -1.18e+01 -1.18e+01 -6.31e+01 -3.43e+01 -3.43e+01 -2.76e+01  0.00e+00
 3.24e+02 -1.18e+00 -2.91e+01  1.27e-01  2.83e+00 -9.68e+02 -6.78e+02 -6.78e+02  0.00e+00 -1.29e+02
-1.27e+02  4.62e-01  1.14e+01 -1.04e+00  1.31e+01  3.80e+02  2.66e+02  2.66e+02  0.00e+00  1.05e+03
-1.86e+02  6.75e-01  1.67e+01  8.61e-01 -1.71e+01  5.56e+02  3.89e+02  3.89e+02  0.00e+00 -8.75e+02
 3.42e+02  1.09e+00  1.05e+03  7.56e+02  7.57e+02 -2.98e+01  1.65e-01  3.28e+00  0.00e+00  0.00e+00
-3.08e+01 -9.82e-02 -9.47e+01 -6.80e+01 -6.80e+01  2.68e+00 -2.66e+00  4.88e+00  0.00e+00  0.00e+00
-3.02e+02 -9.65e-01 -9.31e+02 -6.69e+02 -6.69e+02  2.63e+01  2.42e+00 -9.56e+00  0.00e+00  0.00e+00
 0.00e+00  0.00e+00  0.00e+00  0.00e+00  0.00e+00  0.00e+00  0.00e+00  0.00e+00 -1.67e+00  0.00e+00
 0.00e+00  0.00e+00  0.00e+00  0.00e+00  0.00e+00  0.00e+00  0.00e+00  0.00e+00  0.00e+00 -1.00e+01
```

B

```
0.00e+00  0.00e+00
0.00e+00  0.00e+00
0.00e+00  0.00e+00
0.00e+00  0.00e+00
0.00e+00  0.00e+00
0.00e+00  0.00e+00
0.00e+00  0.00e+00
0.00e+00  0.00e+00
1.67e+00  0.00e+00
0.00e+00  1.00e-02
```

C

```
 1.00e+00  0.00e+00  0.00e+00  0.00e+00  0.00e+00  0.00e+00  0.00e+00  0.00e+00  0.00e+00  0.00e+00
-4.91e-01  0.00e+00 -6.32e-01  0.00e+00  0.00e+00 -2.07e-01  0.00e+00  0.00e+00  0.00e+00  0.00e+00
```

APPENDIX C

LINEAR MODELS OF A TYPICAL V/STOL AIRCRAFT

A (COND 1)

```
0.00e+00  1.00e+00  0.00e+00  0.00e+00  0.00e+00  0.00e+00  0.00e+00  0.00e+00  0.00e+00  0.00e+00
-4.16e-06 -5.88e-02 -9.26e-03  1.48e-01 -6.74e+00  0.00e+00 -4.02e-02  5.14e+01 -8.53e+01 -1.45e+01
-5.56e-01 -2.05e-02 -2.88e-02  4.12e-04  4.67e-02  0.00e+00 -5.63e-01  8.48e+00  1.44e+00  1.09e+00
-7.81e-02  1.49e-01 -1.80e-03 -4.00e-02  1.20e-01  0.00e+00 -6.31e-02 -5.91e+01 -9.67e+00 -7.40e+00
 0.00e+00  0.00e+00  0.00e+00  0.00e+00 -2.00e+01  0.00e+00  0.00e+00  0.00e+00  0.00e+00  0.00e+00
 0.00e+00  0.00e+00  0.00e+00  0.00e+00  0.00e+00 -1.00e+01  0.00e+00  0.00e+00  0.00e+00  0.00e+00
 0.00e+00  0.00e+00  0.00e+00  0.00e+00  0.00e+00  0.00e+00 -5.00e+00  0.00e+00  0.00e+00  0.00e+00
 0.00e+00  0.00e+00  1.09e-05  4.34e-07  0.00e+00  0.00e+00  0.00e+00 -4.17e+00  2.28e+00  9.16e-01
 0.00e+00  0.00e+00  5.99e-05  7.05e-06  0.00e+00  0.00e+00  0.00e+00  1.01e-04 -4.05e+00  7.86e-01
 0.00e+00  0.00e+00 -7.95e-04 -9.34e-05  0.00e+00  2.98e+01  0.00e+00 -5.82e+01  0.00e+00 -1.33e+01
```

A (COND 2)

```
0.00e+00  1.00e+00  0.00e+00  0.00e+00  0.00e+00  0.00e+00  0.00e+00  0.00e+00  0.00e+00  0.00e+00
-4.02e-06 -4.22e-01  1.26e-01 -1.58e-02 -6.80e+00  0.00e+00  2.03e-02  5.83e+01 -8.53e+01 -1.43e+01
-5.56e-01 -2.84e-01 -4.24e-02 -2.30e-02 -5.55e-02  0.00e+00 -4.75e-01  1.69e+01  3.21e+00  2.00e+00
-7.81e-02  2.03e+00 -4.85e-02 -2.04e-01 -2.10e-01  0.00e+00 -1.36e-01 -6.82e+01 -1.05e-01 -6.60e+00
 0.00e+00  0.00e+00  0.00e+00  0.00e+00 -2.00e+01  0.00e+00  0.00e+00  0.00e+00  0.00e+00  0.00e+00
 0.00e+00  0.00e+00  0.00e+00  0.00e+00  0.00e+00 -1.00e+01  0.00e+00  0.00e+00  0.00e+00  0.00e+00
 0.00e+00  0.00e+00  0.00e+00  0.00e+00  0.00e+00  0.00e+00 -5.00e+00  0.00e+00  0.00e+00  0.00e+00
 0.00e+00  0.00e+00  1.87e-05  1.69e-06  0.00e+00  0.00e+00  0.00e+00 -4.57e+00  2.41e+00  1.01e+00
 0.00e+00  0.00e+00  3.27e-05  4.05e-06  0.00e+00  0.00e+00  0.00e+00 -1.96e-04 -3.91e+00  7.96e-01
 0.00e+00  0.00e+00  2.01e-06  3.58e-05  0.00e+00  2.52e+01  0.00e+00 -5.83e+01  0.00e+00 -1.33e+01
```

A (COND 3)

```
0.00e+00  1.00e+00  0.00e+00  0.00e+00  0.00e+00  0.00e+00  0.00e+00  0.00e+00  0.00e+00  0.00e+00
-3.98e-06 -6.92e-01  1.69e-01 -1.81e-02 -7.34e+00  0.00e+00  1.58e-01  6.29e+01 -6.64e+01 -1.20e+01
-5.56e-01 -4.87e-01 -4.16e-02 -8.25e-02 -4.25e-02  0.00e+00 -3.74e-01  1.46e+01  5.68e+00  2.95e+00
-7.81e-02  3.47e+00  1.80e-02 -2.95e-01 -4.09e-01  0.00e+00 -1.20e-01 -2.91e+01 -7.18e+00 -3.90e+00
 0.00e+00  0.00e+00  0.00e+00  0.00e+00 -2.00e+01  0.00e+00  0.00e+00  0.00e+00  0.00e+00  0.00e+00
 0.00e+00  0.00e+00  0.00e+00  0.00e+00  0.00e+00 -1.00e+01  0.00e+00  0.00e+00  0.00e+00  0.00e+00
 0.00e+00  0.00e+00  0.00e+00  0.00e+00  0.00e+00  0.00e+00 -5.00e+00  0.00e+00  0.00e+00  0.00e+00
 0.00e+00  0.00e+00 -1.13e-05 -1.56e-06  0.00e+00  0.00e+00  0.00e+00 -3.75e+00  2.54e+00  1.12e+00
 0.00e+00  0.00e+00 -1.19e-05 -1.45e-06  0.00e+00  0.00e+00  0.00e+00 -6.28e-04 -2.71e+00  8.22e-01
 0.00e+00  0.00e+00  8.21e-04  8.74e-05  0.00e+00  1.56e+01  0.00e+00 -5.82e+01  0.00e+00 -1.33e+01
```

A (COND 4)

```
0.00e+00  1.00e+00  0.00e+00  0.00e+00  0.00e+00  0.00e+00  0.00e+00  0.00e+00  0.00e+00  0.00e+00
-3.53e-06 -8.50e-01 -7.46e-02 -2.60e-01 -5.92e+00  0.00e+00  6.76e-01  3.22e+01 -2.51e-01  1.99e+00
-5.56e-01 -6.49e-01 -2.99e-02 -8.48e-02  9.34e-02  0.00e+00 -8.92e-02  1.48e+01  1.59e+01  4.99e+00
-7.81e-02  4.62e+00 -1.67e-01 -4.86e-01 -3.90e-01  0.00e+00 -1.08e-01 -3.79e+00 -2.23e+00 -7.22e-01
 0.00e+00  0.00e+00  0.00e+00  0.00e+00 -2.00e+01  0.00e+00  0.00e+00  0.00e+00  0.00e+00  0.00e+00
 0.00e+00  0.00e+00  0.00e+00  0.00e+00  0.00e+00 -1.00e+01  0.00e+00  0.00e+00  0.00e+00  0.00e+00
 0.00e+00  0.00e+00  0.00e+00  0.00e+00  0.00e+00  0.00e+00 -5.00e+00  0.00e+00  0.00e+00  0.00e+00
 0.00e+00  0.00e+00  1.15e-05  1.74e-06  0.00e+00  0.00e+00  0.00e+00 -2.71e+00  2.62e+00  1.44e+00
 0.00e+00  0.00e+00  2.42e-06  6.19e-07  0.00e+00  0.00e+00  0.00e+00 -6.75e-04 -1.53e+00  9.58e-01
 0.00e+00  0.00e+00  0.00e+00  0.00e+00  0.00e+00  2.01e+01  0.00e+00  0.00e+00  0.00e+00 -1.33e+01
```

A   (COND 5)

```
0.00e+00  1.00e+00  0.00e+00  0.00e+00  0.00e+00  0.00e+00  0.00e+00  0.00e+00  0.00e+00  0.00e+00
-4.71e-06 -1.03e+00  3.48e-02 -4.36e-01 -8.06e+00  0.00e+00  0.00e+00  2.98e+00 -1.52e+00 -2.56e-01
-5.60e-01 -3.81e-01 -4.39e-02  4.87e-02  4.55e-02  0.00e+00  0.00e+00  2.18e+01  1.55e+01  5.21e+00
-3.66e-02  5.82e+00 -1.54e-01 -5.50e-01 -6.95e-01  0.00e+00  0.00e+00 -1.26e+00 -4.05e-01 -1.36e-01
 0.00e+00  0.00e+00  0.00e+00  0.00e+00 -2.00e+01  0.00e+00  0.00e+00  0.00e+00  0.00e+00  0.00e+00
 0.00e+00  0.00e+00  0.00e+00  0.00e+00  0.00e+00 -1.00e+01  0.00e+00  0.00e+00  0.00e+00  0.00e+00
 0.00e+00  0.00e+00  0.00e+00  0.00e+00  0.00e+00  0.00e+00 -5.00e+00  0.00e+00  0.00e+00  0.00e+00
 0.00e+00  0.00e+00  2.80e-05  1.83e-06  0.00e+00  0.00e+00  0.00e+00 -2.74e+00  2.61e+00  1.45e+00
 0.00e+00  0.00e+00  6.97e-06  3.90e-07  0.00e+00  0.00e+00  0.00e+00 -1.03e-03 -1.60e+00  9.82e-01
 0.00e+00  0.00e+00  0.00e+00  0.00e+00  0.00e+00  2.01e+01  0.00e+00  0.00e+00  0.00e+00 -1.33e+01
```

A   (COND 6)

```
 0.00e+00  1.00e+00  0.00e+00  0.00e+00  0.00e+00  0.00e+00  0.00e+00  0.00e+00  0.00e+00  0.00e+00
 1.60e-06 -1.29e+00  8.11e-03 -5.61e-01 -1.26e+01  0.00e+00  0.00e+00 -2.21e+00 -1.75e+00 -3.23e-01
-5.61e-01  2.00e-02 -5.14e-02  1.21e-02 -3.00e-03  0.00e+00  0.00e+00  8.32e+00  1.89e+01  5.96e+00
 1.53e-03  7.29e+00 -1.53e-01 -5.78e-01 -1.09e+00  0.00e+00  0.00e+00 -5.44e-01 -4.94e-01 -1.56e-01
 0.00e+00  0.00e+00  0.00e+00  0.00e+00 -2.00e+01  0.00e+00  0.00e+00  0.00e+00  0.00e+00  0.00e+00
 0.00e+00  0.00e+00  0.00e+00  0.00e+00  0.00e+00 -1.00e+01  0.00e+00  0.00e+00  0.00e+00  0.00e+00
 0.00e+00  0.00e+00  0.00e+00  0.00e+00  0.00e+00  0.00e+00 -5.00e+00  0.00e+00  0.00e+00  0.00e+00
 0.00e+00  0.00e+00  4.41e-07  9.72e-07  0.00e+00  0.00e+00  0.00e+00 -2.82e+00  2.82e+00  1.39e+00
 0.00e+00  0.00e+00 -1.68e-05  9.73e-08  0.00e+00  0.00e+00  0.00e+00 -1.84e-03 -1.53e+00  9.05e-01
 0.00e+00  0.00e+00  0.00e+00  0.00e+00  0.00e+00  2.01e-01  0.00e+00  0.00e+00  0.00e+00 -1.33e+01
```

A   (COND 7)

```
 0.00e+00  1.00e+00  0.00e+00  0.00e+00  0.00e+00  0.00e+00  0.00e+00  0.00e+00  0.00e+00  0.00e+00
 1.53e-05 -1.54e+00 -8.81e-03 -6.80e-01 -1.82e+01  0.00e+00  0.00e+00 -6.77e+00 -2.27e+00 -3.73e-01
-5.61e-01  3.48e-01 -6.26e-02 -1.67e-02 -6.25e-02  0.00e+00  0.00e+00  5.36e+01  1.70e+01  6.57e+00
 2.21e-02  6.75e+00 -1.36e-01 -8.09e-01 -1.57e+00  0.00e+00  0.00e+00 -1.17e+00 -4.48e-01 -1.72e-01
 0.00e+00  0.00e+00  0.00e+00  0.00e+00 -2.00e+01  0.00e+00  0.00e+00  0.00e+00  0.00e+00  0.00e+00
 0.00e+00  0.00e+00  0.00e+00  0.00e+00  0.00e+00 -1.00e+01  0.00e+00  0.00e+00  0.00e+00  0.00e+00
 0.00e+00  0.00e+00  0.00e+00  0.00e+00  0.00e+00  0.00e+00 -5.00e+00  0.00e+00  0.00e+00  0.00e+00
 0.00e+00  0.00e+00  1.01e-04 -4.86e-06  0.00e+00  0.00e+00  0.00e+00 -4.46e+00  2.87e+00  1.26e+00
 0.00e+00  0.00e+00 -9.47e-06 -3.21e-07  0.00e+00  0.00e+00  0.00e+00 -3.17e-03 -3.58e+00  8.51e-01
 0.00e+00  0.00e+00  0.00e+00  0.00e+00  0.00e+00  4.11e+01  0.00e+00  0.00e+00  0.00e+00 -1.33e+01
```

B   (COND 1,2,3,4)

```
0.00e+00  0.00e+00  0.00e+00
0.00e+00  0.00e+00  0.00e+00
0.00e+00  0.00e+00  0.00e+00
0.00e+00  0.00e+00  0.00e+00
2.00e+01  0.00e+00  0.00e+00
0.00e+00  1.00e+01  0.00e+00
0.00e+00  0.00e+00  5.00e+00
0.00e+00  0.00e+00  0.00e+00
0.00e+00  0.00e+00  0.00e+00
0.00e+00  0.00e+00  0.00e+00
```

B - (COND 5,6,7)

```
0.00e+00  0.00e+00
0.00e+00  0.00e+00
0.00e+00  0.00e+00
0.00e+00  0.00e+00
2.00e+01  0.00e+00
0.00e+00  1.00e+01
0.00e+00  0.00e+00
0.00e+00  0.00e+00
0.00e+00  0.00e+00
0.00e+00  0.00e+00
```

C.   (COND $1, 2, ..., 7$)

```
[ 2.00e-01  0.00e+00  0.00e+00  0.00e+00  0.00e+00  0.00e+00  0.00e+00  0.00e+00  0.00e+00  0.00e+00
  0.00e+00  0.00e+00  4.00e-02  0.00e+00  0.00e+00  0.00e+00  0.00e+00  0.00e+00  0.00e+00  C.00e+00
  0.00e+00  0.00e+00  0.00e+00  1.00e-01  0.00e+00  0.00e+00  0.00e+00  0.00e+00  0.00e+00  0.00e+00 ]
```

Note

The A matrices have as a common structure

$$A = \left(\begin{array}{ccc} A_1 & \vdots & \left(\begin{array}{ccc} A_2 & 0 & A_3 \end{array} \right) & \vdots & A_4 \\ \hline 0 & \vdots & \left(\begin{array}{ccc} a_{5,5} & 0 & 0 \\ 0 & a_{6,6} & 0 \\ 0 & 0 & a_{7,7} \end{array} \right) & \vdots & 0 \\ \hline A_5 & \vdots & \left(\begin{array}{ccc} 0 & 0 & 0 \\ 0 & 0 & 0 \\ 0 & a_{10,6} & 0 \end{array} \right) & \vdots & A_6 \end{array} \right).$$

$A_1 \in \Re^{4 \times 4}$ and $A_6 \in \Re^{3 \times 3}$ are the cross-coupling matrices for the aerodynamic states (X1 - X4) and the engine states (X8 - X10) respectively (fig. 9.14). States X5, X6 and X7 are the actuator states and $a_{5,5}$, $a_{6,6}$ and $a_{7,7}$ are their respective time constants. The aerodynamic states are affected by the actuators states X4 (elevator angle) and X6 (nozzle angle) through the feedforward gain vectors A_2 and A_3, while the engine states are affected by the actuator state X5 (throttle) through the gain $a_{10,6}$. Also, the cross-coupling from the engine states to the aerodynamic ones ($A_4 \in \Re^{4 \times 3}$) is significant while that in reverse is negligible ($A_5 \in \Re^{3 \times 4} \approx 0$).

REFERENCES

[Bad 87] Bada, A.T., 1987, "Robust Brake Control for a Heavy-duty Truck", *Proc. IEE*, Vol. 134, Pt. D, No. 1, pp. 1-8.

[Bog 86] Bogetoft, P., 1986, "General Communication Schemes for Multiple Criteria Decision Making", *European J. Operational Research*, Vol. 26, pp. 106-122.

[Box 65] Box, M.J., 1965, "A New Method of Constrained Optimization and Comparison with Other Methods", *Computer J.*, Vol. 8, No. 42, pp. 42-52.

[Boy 88] Boyd, S.P. et al., 1988, "A New CAD Method and Associated Architectures for Linear Controllers", *IEEE Trans. Automatic Control*, Vol. 33, No. 3, pp. 268-283.

[Bra 84] Brans, J.P., Mareschal, B. and Vincke, Ph., 1984, "PROMETHEE - A New Family of Outranking Methods in Multicriteria Analysis" in *Operational Research*, Brans, J.P. (Ed.), Elsevier Science Publishers.

[Bre 88] Breuer, P.T., 1987, "A Data Language - DL", to be published as internal report, Cambridge University Engineering Department, Cambridge, U.K.

[Cha 74] Chamberlin, D.D. and Boyce, R.F., 1974, "SEQUEL: A Structured English Query Language", *Proc. ACM-SIGFIDET Workshop*, Ann Arbor, MI, U.S.A.

[Cha 76] Chamberlin, D.D., et al., 1976, "SEQUEL-2: A Unified Approach to Data Definition, Manipulation, and Control", *IBM J. Res. Develop.*, November, 1976, pp. 560-575.

[Cod 71] Codd, E.F., 1971, "A Relational Model of Data for Large Shared Data Banks", *Commun. ACM*, Vol. 13, pp. 377-387.

[Cod 72] Codd, E.F., 1972, "Relational Completeness of Data Base Sublanguages", in *Data Base Systems*, Courant Computer Science Symposia Series, Vol. 6, Englewood Cliffs, N.T., Prentice Hall.

[DFV 84] DFVLR, 1984, *Institute for Flight System Dynamics (Preprints for Scientific Report of the Research Department)*, Flight Mechanics/Guidance and Control, DFVLR, Oberpfaffenhofen, D-8031, Wessling, Federal Republic of Germany.

[Dat 80] Date, C.J., 1980, *An Introduction to Database System*, Addison-Wesley.

[Den 84]   Denham, M.J., 1984, "Design Issues for CACSD Systems", *Proc. IEEE*, Vol. 72, No. 12, pp. 1714-1723.

[Fle 86]   Fleming, P.J. and Pashkevich, A.P., 1986, "Application of Multi-objective Optimization to Compensator Design for SISO Control Systems", *Electronics Letters*, Vol. 22, No. 5, pp. 258-259.

[Fle 87]   Fletcher, R., 1987, *Practical Methods of Optimization*, John Wiley and Sons.

[Fra 86]   Franklin, G.F., Powell, J.D. and Emami-Naeini, A., 1986, *Feedback Control of Dynamic Systems*, Addison-Wesley.

[Geo 72]   Geoffrion, A.M., Dyer, J.S. and Feinberg, A., 1972, "An Interactive Approach for Multicriterion Optimization", *Management Science*, Vol. 10, pp. 357-368.

[Glo 88]   Glover. K.G. and Doyle, J.C., 1988, "State Space Formulae for All Stabilizing Controllers that Satisfy an H^∞ norm bound and Relations to Risk Sensitivity", to be published in *Systems and Control Letters*.

[Goi 82]   Goicoeche, A.G., Hansen, D.R. and Duckstein, L., 1982, *Multiple Objective Decision Analysis with Engineering and Business Applications*, Wiley, New York.

[Gol 85]   Golpalsami, N. and Sanathanan, C.K., 1985, "Satisfactory Solution s Approach to Parameter Optimization of Dynamical Systems with Vector Performance Index, *J. Opt. Th. and App.*, Vol. 47, No. 3, pp. 301-319.

[Hor 79]   Horowitz, I., 1979, "Quantitative Synthesis of Uncertain Multiple Input-Ouput Feedback System", *Int. J. Control*,, Vol. 30, No. 1, pp. 81-106.

[Jac 81]   Jacob, H.G. and Deng, J., 1981, "Computer-Supported Multiple Objective Pareto Optimal Procedures as Applied to a Flight Control System", *Automatiserungstechnik at*, Vol. 34, Jahrgang, Heft 9, pp. 346-355.

[Joo 86]   Joos, D. and Yang, W., 1986, "Revised User's Guide to REMVG Concerning Flying Quality Criteria and Low Order Models", Report 515-86/1, DFVLR, Oberpfaffenhofen, D-8031, Wessling, Federal Republic of Germany.

[Kae 86]   Kaesbauer, D., 1986, *Robuster Reglerentwurf durch Kontraktion eines Polgebiets*, Dissertation TU Graz, Austria.

[Ker 76]   Kernighan, B.W. and Plauger, P.J., 1976, *Software Tools*, Addison-Wesley.

[Kie 85]   Kiwiel, K.C., 1985, *Methods of Descent for Nondifferentiable Optimization*, Lecture Notes in Mathematics, 1133, Springer Verlag.

[Kok 85]   Kok, M. and Lootsma, F.A., 1985, "Pairwise Comparison Methods in Multiple Objective Programming with Applications in a Long Term Energy Planning Model", *European J. Operational Research*, Vol. 22, No. 1, pp. 44-55.

[Kok 86]   Kok, M., 1986, "The Interface with Decision Makers and Some Experimental Results in Interactive Multiple Objective Programming Methods", *European J. Operational Research*, Vol. 26, pp. 96-107.

[Kre 83]   Kreisselmeier, G. and Steinhauser, R., 1983, "Application of Vector Performance Optimization to a Robust Control Loop Design for a Fighter Aircraft", *Int. J. Control*, Vol. 37, No. 2, pp. 251-284.

[Lim 79]   Limebeer, D.J.N., Harley, R.G., Schuck, S.M., 1979, "Subsynchronous Resonance of the Koeberg Turbo-generators and of a Laboratory System", *Trans. (South Africa) IEE*, vol. 70, pp. 278-297.

[Lim 85]   Limebeer, D.J.N. and Maciejowski, J.M., 1982, "Two Tutorial Examples of Multivariable Control System Design", *Trans. Inst. of Measurement and Control*, Vol. 7, No. 2, pp. 97-107.

[Lit 84]   Little, J.N., Emmami-Naeini, A. and Bangert, S.N., 1984, "Ctrl-C and Matrix Environments for the Computer-Aided Design of Control Systems", *Proc. 6th Int. Conf. on Analysis and Optimization of Systems*, pp. 191-205.

[Lue 73]   Luenberger, D.G., 1973, *Linear and Non-Linear Programming*, Addison-Wesley.

[Mac 80]   MacFarlane, A.G.J. (Ed.), 1980, *Complex Variable Methods for Linear Multivariable Feedback Systems*, Taylor and Francis Limited, London.

[Mac 84]   Maciejowski, J.M., 1984, "Data Structures for Control System Design", presented at EUROCON 84, Brighton, U.K.

[Mac 87]   MacFarlane, A.G.J., Grubel, G. and Ackermann, J., 1987, "Future Design Environment for Control Engineering", *Proc. IFAC 87 World Congress on Automatic Control*, Vol. 8, pp. 235-246.

[Mac 87]   Maciejowski, J.M., 1987, "Queries and Transactions in a Control Engineering Database", presented at the IEE Symposium in CACSD, London, U.K., Nov., 1987.

[Mat 87]   The MathWorks, Inc., 1987, *Pro-Matlab User's Guide*, The MathWorks, Inc., 20, North Main St., Suite 250, Sherborn, MA 01770, U.S.A.

[Maz 87]   Mazeed, M.A., Sunaga, T. and Kondo, E., 1987, "Non-linear Optimization Using Least Square Function and the Complex Method", *the Memoirs of the Faculty of Engineering*, Kyushu University, Vol. 47, No. 3, pp. 197-175.

[Mol 81]   Moler, C.B., 1981, *Matlab User's Guide*, Report TRCS81-1, Department of Computer Science, University of New Mexico, Alberquerque, New Mexico, U.S.A.

[Nak 84]   Nakayama, H., 1984, "Proposal of Satisficing Trade-Off Method for Multiple Objective Programming", *Trans. Soc. Inst. Control Eng.*, Vol. 20, pp. 29-35 (in Japanese).

[Nel 65]   Nelder, J.A. and Mead R., 1965, "A Simplex Method for Function Minimization", *Computer J.*, Vol. 7, pp. 308-313.

[Ng 87]   Ng, W-Y., 1987, "Application of Optimization-based Methods in Control System Design", presented at the IFIP 13th Int. Conf. in System Modeling and Optimization, Tokyo, Japan (to be included in Iri, M. (Ed.), 1988, *System Modeling and Optimization*, Springer-Verlag).

[Ng 88]   Ng, W-Y., 1988, "Interactive Descriptive Graphical Approach to Data Analysis for Trade-off Decisions in Multi-Objective Programming", presented at the 27th Meeting of the European Working Group on Multiple Criteria Decision Aid, Mons, Belgium.

[Ng 88]   Ng, W-Y., 1988, "A Decision Support System for Multi-Objective Design of Practical Controllers", *Proc. of 1988 American Control Conference*, pp. 713-718.

[Nor 86]   Norman, D.A. and Draper, S.W. (Ed.), 1986, *User Centred System Design*, Lawrence Erlbaum Associates.

[Nye 83]   Nye, W.T., 1983, *DELIGHT: An Interactive System for Optimization-based Engineering Design*, Ph.D. Dissertation, Department of Electronic Engineering and Computer Science, Univ. California, Berkeley, California, U.S.A.

[Nye 86]   Nye, W.T. and Tits, A.L., 1986, "An Application-Oriented Optimization-based Methodology for Interactive Design of Engineering Systems", *Int. J. Control*, vol. 43, No. 6, pp. 1693-1721.

[Orl 78]   Orlvosky, S.A., 1978, "Decision-making with a Fuzzy Preference Relation", *Fuzzy Sets and Systems*, Vol. 1, pp. 155-167.

[Pan 86]   Pang, G.K.H. and Boyle, J.M., 1986, "An Expert System for Analytical and Interactive Design of Control Systems", Presented at the 2nd Int. Expert System Conference, London, U.K.

[Pan 87]   Pang, G.K.H. and MacFarlane, A.G.J., 1987, *An Expert System Approach to Computer-Aided Design of Multivariable Systems*, Lecture Notes in Control and Information Sciences, 89, Springer-Verlag.

[Pay 76]   Payne, J.W., 1976, "Task Complexity and Contingent Processing in Decision Making: An Information Search and Protocol Analysis", *Organizational Behaviour and Human Performance*, Vol. 16, pp. 366-387.

[Pha 87]   Phaal, P., *An Object-Oriented Environment for Control System Design*, Ph.D. Dissertation, Cambridge University Engineering Department, Cambridge, U.K.

[Pol 82]   Polak, E. and Wardi, Y., 1982, "A Non-differentiable Optimization Algorithm for the Design of Control Systems Subject to Singular Value Inequalities over a Frequency Range", *Automatica*, Vol. 8, pp. 267-283.

[Pol 84]   Polak, E., Mayne, D.Q. and Stimler, D.M., 1984, "Control System Design via Semi-Infinite Programming: A Review", *Proc. IEE*, Vol. 72, No. 12, pp. 1777-1794.

[Pol 86]   Polak, E. and Stimler, D.M., 1986, "Majorization: A Computational Complexity Reduction Technique in Control System Design", *Proc. 7th Int. Conf. on Analysis and Optimization of Systems*, Springer-Verlag, pp. 54-64.

[Pow 64]   Powell, M.J.D., 1964, "An Efficient Method for Finding the Minimum of a Function of Several Variables Without Calculating Derivatives", *Computer J.*, Vol. 7, pp. 155-162.

[Ros 60]   Rosenbrock, H.H., 1960, "An Automatic Method for Finding the Greatest or Least Value of a Function", *Computer J.*, Vol. 3, pp. 175-184.

[Ros 74]   Rosenbrock, H.H., 1974, *Computer Aided Control System Design*, Academic Press, London.

[Ros 85]   Rosenthal, R.E., 1985, "Concepts, Theory and Techniques - Principles of Multi-objective Optimization", *Decision Sciences*, Vol. 16, pp. 133-152.

[Roy 81]   Roy, B. and Vincke, Ph., 1981, "Multiple Criteria Analysis: Survey and New Directions", *European J. Operational Research*, Vol. 8, pp. 207-218.

[Saw 85]   Sawaragi, Y., Nakayama, H., Tanino, T., 1985, *Theory of Multi-Objective Optimization*, Academic Press, Inc.

[Spr 81]   Spronk, J., 1981, *Interactive Multiple Goal Programming: Application to Financial Planning*, Nijhof, Boston, U.S.A.

[Spr 82]   Sprague, R.H. and Carlson, E.D., 1982, *Building Effective Decision Support Systems*, Englewood Cliffs.

[Suc 87]   Suchman, L., *Plans and Situated Actions: Problems of the Man-Machine Communication*, Cambridge University Press.

[Tab 79]   Tabak, D., Schy, A.A., Giesy, D.P. and Johnson, K.G., 1979, "Application of Multi-objective Optimization in Aircraft Control System Design", *Automatica*, Vol. 15, pp. 595-600.

[Tai 86]   Taiwo, O., 1986, "The Design of Robust Control System for Plants with Recycle", *Int. J. Control*, vol. 432, pp. 671-678.

[Tan 87]   Tanenbaum, A.S., 1987, *Operating Systems*, Prentice Hall International, Inc.

[Tho 85]   Thompson, P.M., 1985, "Program CC: CACSD on the IBM-PC", presented at the 2nd IEEE Control System Society Symposium on CACSD, Santa Barbara, U.S.A..

[Tuf 83]   Tufte, E.R., 1983, *The Visual Display of Quantitative Information*, Graphics Press.

[Van 86]   Van de Vegte, J., 1986, *Feedback Control Systems*, Prentice-Hall.

[Wet 86]   Wette, M. and Laub, A.J., 1986, "Software Practices in CACSD: A Need for Tool-based Systems", presented at the IEEE Symposium of CACSD, Arlington, Virginia, U.S.A.

[Wie 82]   Wierzbicki, A.P., 1982, "A Mathematical Basis for Satisficing Decision Making", *Mathematical Modelling*, Vol. 3, pp. 391-405.

[Wie 85]   Wierzbicki, A.P., 1985, "A Methodological Approach to Composing Parametric Characterizations of Efficient Solutions" in *Large-Scale Modelling and Interactive Decision Analysis*, Fandel et al. (Ed.), Lecture Notes in Economics and Mathematical Systems, 273, Springer-Verlag.

[Win 84]   Winston, P.H., 1984, Artificial Intelligence, Addison-Wesley.

179

[Zak 73]   Zakian, V. and Al-Naib, U. 1973, "Design of Dynamical and Control Systems by the Method of Inequalities", *Proc. IEE*, Vol. 120, pp. 1421-1427.

[Zak 84]   Zakian, V., 1984, "A Framework for Design: Theory of Majorants", Report 604, Control System Centre, UMIST, Manchester, U.K.

[Zel 82]   Zeleny, M., 1982, *Multiple Criteria Decision Making*, McGraw Hill, Inc.

[Zio 83]   Ziont, S. and Wallenius, J., 1983, "An Interactive Multiple Objective Linear Programming Method for a Class of Underlying Nonlinear Utility Function", *Management Science*, Vol. 29, pp. 519-529.

Lecture Notes in Control and Information Sciences

Edited by M. Thoma and A. Wyner

Lecture Notes in Control and Information Sciences

Edited by M. Thoma and A. Wyner

Lecture Notes in Control and Information Sciences

Edited by M. Thoma and A. Wyner